# ILLUSTRATED ENCYCLOPEDIA OF
# ASTRONOMY

# ILLUSTRATED ENCYCLOPEDIA OF
# ASTRONOMY

GENERAL EDITOR
## JOHN MAN

FOREWORD BY
## DR CARL SAGAN

CHANCELLOR
PRESS

# CONTENTS

## CONTENTS

Some material in this book previously appeared in other Hamlyn books. It has been edited and supplemented by Martin Wace using original text by Geoffrey Bath, Heather Couper, Storm Dunlop, Leo Enright, Peter Francis, John Gribbin, Nigel Henbest, Sir Bernard Lovell, Peter Owen and Ian Ridpath.

First published 1989 by The Hamlyn Publishing Group Limited
This edition published in 1996 by Chancellor Press, an imprint of
Reed International Books, Michelin House, 81 Fulham Road, London SW3 6RB
and Auckland, Melbourne, Singapore and Toronto

Copyright © in this edition, Reed International Books 1989

ISBN 1 85153 007 X

Printed in China

# FOREWORD

By Carl Sagan
Director, Laboratory for Planetary Studies
David Duncan Professor of Astronomy and Space Sciences
Cornell University, Ithaca, New York, USA

In the early 17th century the great Italian scientist Galileo Galilei turned the first small astronomical telescope to the heavens and discovered wonders: that Venus underwent phases from new to crescent to full, like the Earth's moon, demonstrating that the Earth went around the Sun and not vice versa; that there were mountains and craters on the Moon. implying that it was a world, in some sense like ours, and not made of some special sky stuff, such as Aristotle's 'quintessence'; that there were dark spots on the Sun's visible disk, showing that it was not 'perfect', as 2000 years of religious mystics had maintained; and that there were four moons orbiting the planet Jupiter, shattering a theological ban on the existence of new worlds. Galileo's work, including his observations of the motions of the moons of Jupiter, was an important springboard for Isaac Newton's invention of mathematical physics, which has in turn led, in an important way, to our modern technological civilization.

Once Galileo had improved Dutch novelty lenses to make the first astronomical telescope, these breathtaking discoveries were inevitable. The astronomical phenomena were up there waiting to be discovered; it only required a fairly minor technological improvement over our human eyesight. Today a comparable revolution is being worked in astronomy. We are no longer restricted to ordinary visible light. Radio waves emitted by many astronomical objects are transmitted through the Earth's atmosphere and are now being studied by immense radio telescopes on the Earth's surface. The Earth's atmosphere is entirely opaque to gamma rays and x-rays, and partially opaque to ultraviolet and infrared light. Accordingly, we have launched instruments sensitive to these wavelengths to high altitudes, above much or all of the Earth's atmosphere – in airplanes, in balloons, in rockets and particularly in Earth-orbiting satellites. Until recently astronomy has been a kind of passive science in which the practitioners wistfully scan the heavens hoping something would happen. But now we have radar astronomy which is able to perform remote experiments on nearby objects; and, of the greatest significance, we have begun launching small instrumented space vehicles to fly by, orbit, and land on the nearest celestial objects. We have examined closely all of the planets known to the ancients. And we have sent four spacecraft – Pioneers 10 and 11 and Voyagers 1 and 2 – on trajectories which will ultimately take them out of the solar system altogether, and into the realm of the stars.

With such remarkable instruments the pace of astronomical discovery has in the last decade or two become breathtaking. In many areas it has moved from myth and vague guesswork to precise knowledge. At the same time many more mysteries have been uncovered as old ones have been resolved. In recent years, the finding of deuterium between the stars with an orbiting ultraviolet observatory has cast light on the evolution of the universe: active volcanoes have been found on Io, the innermost of the four moons of Jupiter found by Galileo; distant astronomical objects have been found with apparent velocities greater than that of light, providing an interesting challenge to the theoreticians; the first serious searches in human history have been made for simple life on the planet Mars and for advanced civilizations on the planets of other stars; high energy cosmic rays discovered on the Earth have been identified as arising from colossal explosions in remote galaxies; ring systems have been identified around Jupiter and Uranus; a new worldlet has been discovered beyond the orbit of Saturn, and Pluto has been found to have a massive moon; radio observations reveal that two orbiting pulsars behave precisely as predicted by Einstein's theory of general relativity; and evidence has been presented suggesting that the clouds of Titan (the great moon of Saturn) and the grains and gas between the stars are all composed in part of organic molecules, related to those which four billion years ago, on our planet, led to the origin of life.

Like Galileo's findings, these recent discoveries, fragmentary data, and provocative first attempts have a significance ranging far beyond the perview of specialists in astronomy. We are in the midst of discovering nothing less than the nature of the universe and our place in it. In the long run some of these findings will have the most profound practical, as well as philosophical consequences. Modern astronomy and space sciences are providing answers to questions which we have asked for as long as there have been human beings. It evokes our sense of wonder; it speaks to us of who we are.

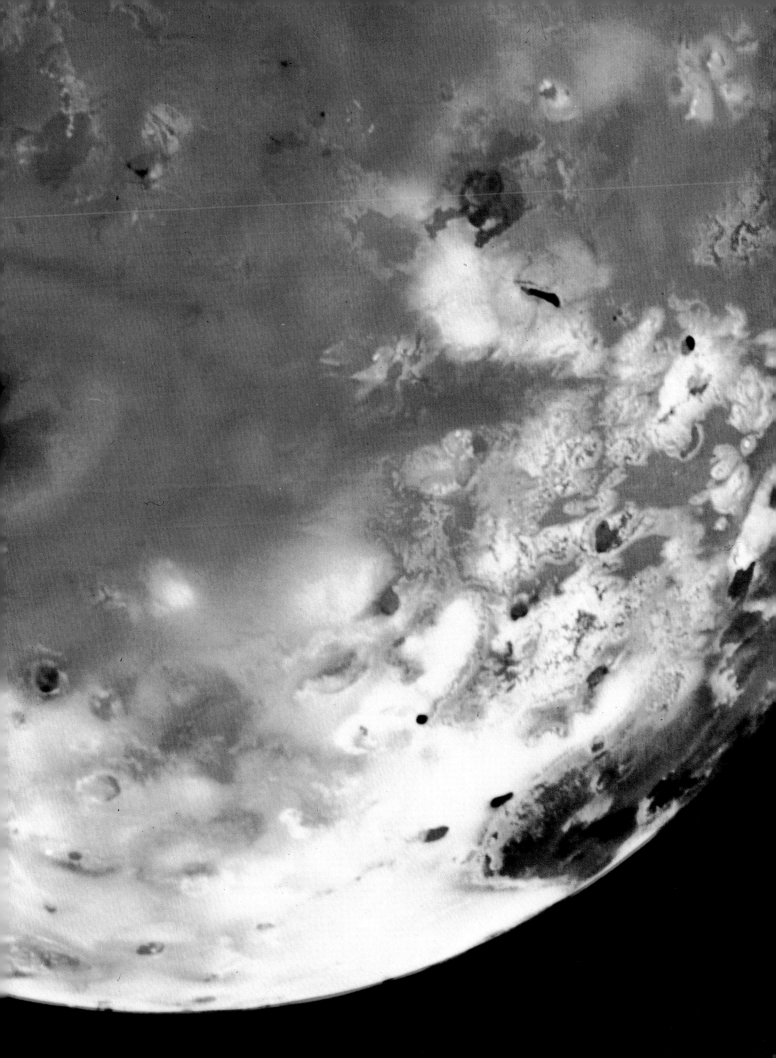

# 1/INTRODUCTION

We are at present in the midst of a revolution in astronomy as profound in its way as that initiated by Copernicus when he showed that the Sun, not the Earth, is at the centre of our Solar System. At one level, the revolution is on the grandest of all themes – the beginning and end of the Universe. In addition, new space technologies will, within a few decades, allow mankind to live permanently in space. *Homo sapiens* may become a galactic, rather than simply a terrestrial species. This opening chapter looks at our place in the Universe and puts the importance of our Solar System in its true perspective.

Is it possible to comprehend the scale of the Universe? It's easy to describe the Universe in conventional astronomical terms, but the question of *comprehending* it is an entirely different matter. It is a very long time since anyone has been able to comprehend the size of the Universe. 2000 years ago, Posidonius calculated the distance of the Sun as 6500 Earth diameters. It wasn't a bad estimate: he was only out by a factor of two and it was a distance that could be comprehended if one had a feeling for the size of the Earth as the ancient Greeks did. Copernicus used a complicated argument to prove that the stars, which he assumed were all the same distance, were fixed to a sphere one and a half million Earth radii distance – a long way, but still comprehensible. But in 1838, when Friedrich Bessel measured the distance to a nearby star, 61 Cygni, he obtained an answer of about ten light-years, and today we know that the nearest stars are more than 5000 million Earth radii distant.

These immense distances are entirely incomprehensible in everyday terms. If we try to think of such distances in terms of kilometres or miles, the figures become meaningless. Even comparisons are not much use. If, in your mind, you reduce the Sun to the size of a 2 metre (6 feet) ball, then Pluto, the most distant planet of our solar system, would be a marble 8 km (5 miles) away. But the nearest star would be 55,000 km (34,000 miles) away. And our own galaxy, the Milky Way – our own collection of stars – would be 1300 *million* km (800 million miles) across. So even on that scale, distances soon become meaningless.

The best one can do is to think in terms of the speed of light, 300,000 km (186,000 miles) per second. That is seven times round the world in a second. The light from the Sun takes about eight minutes to reach us. It takes eleven hours to cross our Solar System. The light reaching us now from the nearest star set off four years ago. Our own Galaxy is about a hundred thousand light-years across. And our nearest galactic neighbour, the great spiral in Andromeda (NGC 224) is some two million light-years away. To talk in terms of light-years makes it possible for astronomers to deal with numbers they can handle. But in human terms, the distances are still unimaginably huge. At present rocket speeds, it would take rather more than a human lifetime to get to the nearest star and back again.

We measure extreme distances in terms of what is known as the redshift – distant objects are receding as the whole Universe expands and the light-waves from them are stretched so that when their light is analysed, it seems to be shifted towards the red end of the spectrum. When Edwin Hubble first published his measurements on speeds of recession in the late 1920s, he said that speeds increased by 530 km per second for every 3.25 million light-years of distance. He claimed an accuracy to within 15 per cent. When Allan Sandage estimated the same constant a few years ago, he gave the recession speed as 50 km per second per 3.25 million light-years, just a tenth of

what Hubble gave, and he was still claiming an accuracy of plus or minus 15 per cent. It's far more important, for the history of the Universe, to estimate not how far in terms of kilometres we can see into space, but how far back into time, bearing in mind that the light from these distant objects set off billions of years ago.

## Cosmic fireball

Probably the most significant discovery in recent cosmological research is that of a low level background radiation which fills the whole Universe uniformly. This seems to be the relic of the radiation left over from the early stages of the expansion of our Universe. Further evidence for an expansion of the Universe comes from the observed recession of clusters of distant galaxies. Of course the fact that the Universe is expanding implies that, at some time in the past, all the objects in it were closer together. We now have very clear evidence that the Universe we know today began as a super-dense, super-hot "singularity". There is no escape, apparently, from the fact that about 10,000 million years ago the Universe was in a completely different state.

It is an extraordinary situation. The whole structure of modern mathematics and physics is built upon laws of nature now in operation. Yet we must now confront the possibility that these "laws" simply do not apply over an infinite period of time.

There is some dispute as to the exact time of the origin of this background radiation, but the most likely theory is that it originated at a stage somewhat less than a million years after the beginning of the expansion of the Universe, an event commonly described as the "primordial fireball". If this is correct, we are receiving information that takes us 99.9 per cent of the way back in time to the beginning of the Universe (whereas investigations of remote objects give us nothing like this penetration – they take us back no more than about 75 per cent of the time to the beginning of the expansion). We can explain the events that probably led up to the formation of the background radiation, events which take us back to a minute fraction of a second ($10^{-43}$ or one over ten followed by 42 zeros) after *Time Zero*. But if you ask what was the situation in the $10^{-43}$ seconds after the expansion began, no prediction is possible on the basis of modern theory. We're concerned there with a Universe less than $10^{-33}$ centimetres across, of an incredibly high density. The theories of gravitation and atomic physics break down. We do not know whether this is a fundamental barrier to scientific description or whether there are other theories yet to be discovered.

## Why is the Universe the way it is?

There are many strange features when one searches for the meaning of the history of the early Universe. The primordial fireball contained the reactions which led to the current proportions of hydrogen and helium – 75 per cent and 25 per cent respectively – a balance that explains the evolution of stars. Very small changes in the primordial fireball would have had an immense effect on the Universe. If certain atomic forces had been only slightly greater, then all the hydrogen would have become an isotope of helium and no long-lived stars could possibly exist. Stars would have formed, but they would have used up all their energy reserves in a very short time. There would be no stars like the Sun, which give an output of energy for thousands of millions of years. Only with stability on this time scale can life evolve. If things

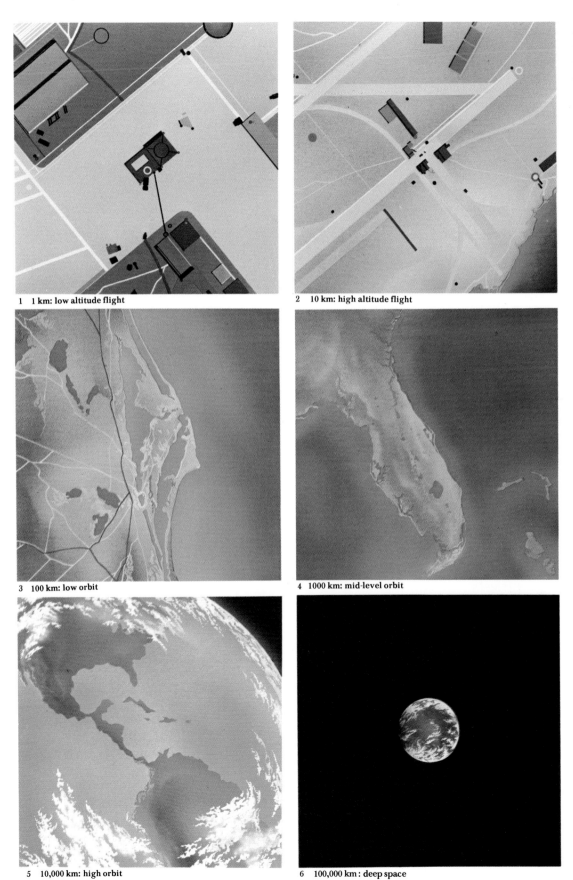

1   1 km: low altitude flight

2   10 km: high altitude flight

3   100 km: low orbit

4   1000 km: mid-level orbit

5   10,000 km: high orbit

6   100,000 km: deep space

The 24 diagrams on this and the following three pages establish the scale of the Universe. Each picture represents a ten-fold increase in scale, from a height of 1 km. (0.6 miles) out to the limits of the known Universe at 10,000 million light-years. The 100,000-fold increase in distance on this page leaps from a bird's eye view of a launch site in the Kennedy Space Center, Cape Canaveral, Florida, to a view of the Earth familiar from Apollo space missions.

had been just a little bit different at the beginning, there would have been no life.

An extraordinary feature pointed out in 1973 by Professor Stephen Hawking of Cambridge University is that, if the early phase of the expansion of the Universe had differed by only one part of a million millionth from what it actually was, there would be no possibility of the Universe existing as we know it. Had the Universe expanded one million millionth part faster, then all its material

These drawings together represent a one million-fold increase in distance over the last picture on the previous page. The Earth-Moon system recedes rapidly into insignificance. In the final two pictures, the outer ring is the orbit of Pluto, which is shown at its average distance – it is in fact an irregular orbit that swings inside that of Neptune. At these scales the Sun becomes a mere pin-prick of light.

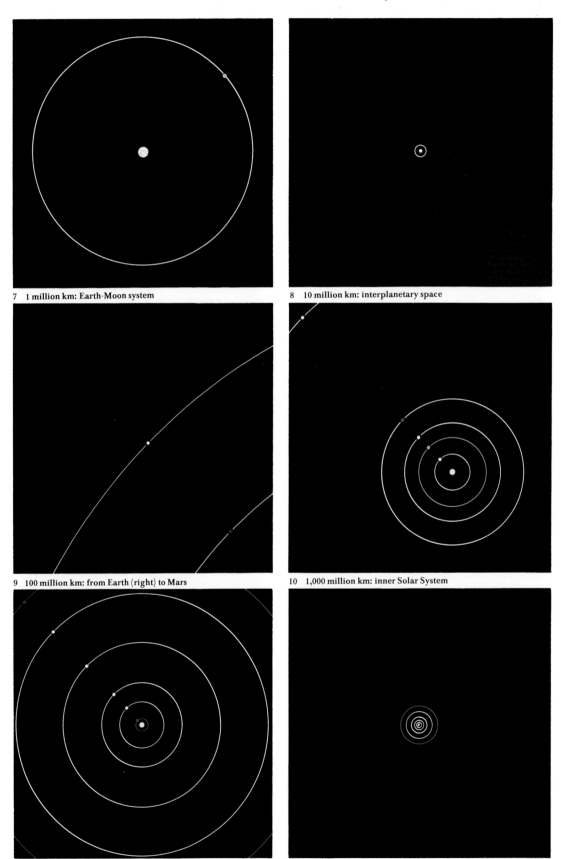

7   1 million km: Earth-Moon system

8   10 million km: interplanetary space

9   100 million km: from Earth (right) to Mars

10   1,000 million km: inner Solar System

11   10,000 million km: Solar System

12   100,000 million km: interstellar space

would have dispersed by now, there would have been no possibility of the gas being drawn together by gravity into stars. If it had been a million millionth part slower, gravitational forces would have caused the Universe to collapse within the first thousand million years of its existence. Again there would have been no long-lived stars and no life.

Astronomers have extended the borders of science to a point where science is no longer

In the depths of interstellar space, the Sun remains a lonely point of light until a 100-fold increase in scale reveals the presence of its nearest neighbours. Distances can now not easily be measured in kilometres, and light-years become a more convenient scale. With further increases in scale, the Sun is lost among its neighbours, whose images begin to overlap as the larger structure that binds them, the Milky Way Galaxy with its characteristic Catherine Wheel spiral, begins to come into view.

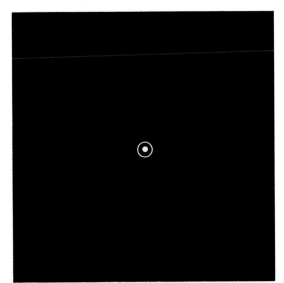

13    1 million million km: interstellar space

14    10 million million km (1 light year): interstellar space

15    10 light years: local stars

16    100 light years: local stars

17    1,000 light years: local region of galaxy

18    10,000 light years: arms of galactic spiral

capable of answering all the questions raised. If at some point in the past, the Universe was once close to a singular state of infinitely small size and infinite density, we have to ask what was there before and what was outside the Universe. The answer at this moment is that such questions have no meaning. Ask a modern astronomer what was outside the Universe before Time Zero. The only possible answer they will be able to give is that there was no "outside".

The Galactic spiral with its 100 billion stars is just one of an uncounted number of galaxies, the nearest of which is two million light-years away. The galaxies themselves are grouped into clusters. These vast systems are all moving away from each other at a speed that is proportional to their distance apart. Light from the most distant known galaxies takes 10,000 million years to reach the Earth, and information about objects at such great distances is scanty, but astronomers have not yet seen any sign that the galaxies thin out.

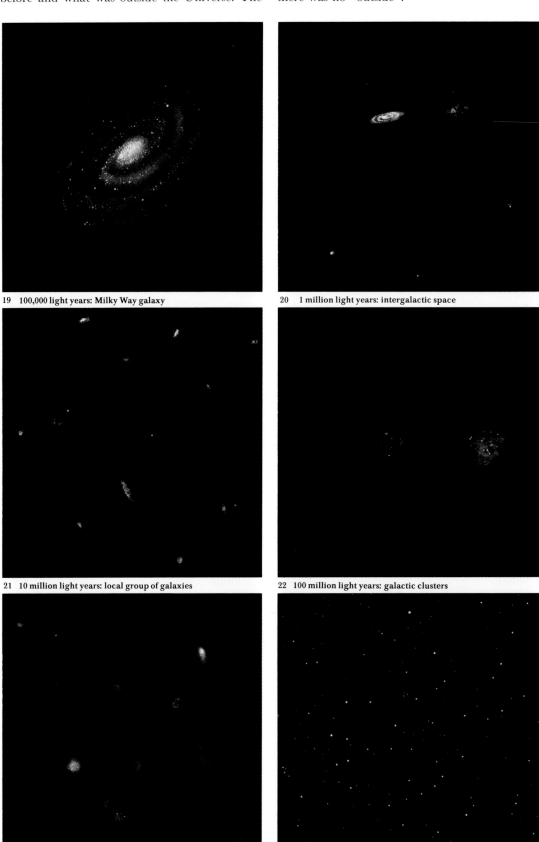

19   100,000 light years: Milky Way galaxy

20   1 million light years: intergalactic space

21   10 million light years: local group of galaxies

22   100 million light years: galactic clusters

23   1,000 million light years: galactic clusters

24   10,000 million light years: limits of known universe

## On the major problems of astronomy

What is the future of the Universe and what is the nature of gravitation? Will the Universe expand for ever, or will it collapse again? The answer depends critically on the amount of matter in the Universe. If the density is at present more than $10^{-29}$ grammes per cc, then we calculate that there is enough mass in the Universe to overcome the present expansion, and the Universe will eventually collapse to another state of super-density. If the density is less than that value the Universe will continue to expand for ever.

The issue is of immense interest. If this is a once-and-for-all Universe, the problems of the Beginning may be insoluble. If it is a cyclical series of many Universes, then it is possible to say, as some cosmologists do, that we are living in one of an infinite cycle of Universes. If so, we are in a privileged position, because this particular Universe has the combination of atomic and gravitational constants such that long-lived stars can exist, which, in turn, make possible the development of life.

What then, of gravity? It is a most remarkable fact that the major force which controls our lives and which determines the large-scale structure of the Universe is not understood. Although Newton discovered the laws of motion and the gravitational law, he evaded the problem of the nature of gravitation. He regarded gravity as a force of unknown type acting at a distance and transmitted instantaneously. Since the publication of Einstein's general theory of relativity in 1915 we have been forced to abandon Newton's concept of an absolute space and time and to regard the gravitational forces as a function of the geometry of space. These forces can no longer be envisaged as acting instantaneously at a distance but are propagated with the speed of light as gravitational waves. But no one has yet proved unequivocally the existence of gravitational waves. We do not understand the relationship between gravitational forces and the other forces which govern the structure and behaviour of atoms.

Although we think we have an accurate description of the Universe, there are great and unsolved problems. Take for instance the problem of black holes, bodies with gravitational fields so intense that not even light can escape from them. Observations have been made that provide strong evidence of their existence, but many astronomers express great caution on the subject. It may be that when the problem of the nature of gravity is solved, then the problem of whether black holes exist or not may completely disappear.

It's worth emphasizing that some very fundamental assumptions about modern astronomy are not universally accepted. It's *generally* accepted, for instance, that the observed redshift is a proof that the Universe is expanding. But not *everybody* believes this. Some astronomers preserve an open mind on this question, and wonder whether the redshift may not have some other explanation, such that the atomic and gravitational constants vary with time. It may be that the existence of the background radiation itself is not incontrovertible evidence for the expansion of the Universe. But new discoveries may yet prove us wrong.

## Is there life elsewhere?

Around twenty years ago, it was a widely held opinion that the development of organic material would have occurred as part of the normal process of the evolution of a typical planetary system, whether our solar system or any other. It was certainly the case that a number of strong arguments for the general existence of life, could hold sway over the doubters. Molecular astronomy had shown the existence of a large variety of quite complex molecules in interstellar space. Additionally, there are good reasons to believe that planetary systems form as a natural byproduct of star formation, and results from the Infrared Astronomical Satellite (IRAS) tend to support this belief.

Yet there are also some interesting arguments against life being common in the universe. It may be that life can only arise and evolve under earth-like conditions. Further, it may be that such conditions are extraordinarily rare, if not unique.

Theoretical arguments can show that, around 4500 million years ago, Earth and Venus were probably very similar. Yet now Venus has a surface temperature that will melt lead and an atmosphere mainly of carbon dioxide with sulphuric acid rain. Where Earth's atmosphere is benign and supportive of life, that of Venus is totally hostile and positively detrimental to any form of life that may have once begun to exist.

The atmosphere of Earth, unlike those of the outer planets, is radically different from the composition of the solar nebula from which the planets formed. The terrestrial atmosphere is a secondary one, probably formed by volcanic eruption. Some unknown cataclismic accident swept away the primordial atmosphere of the inner planets in the very early history of the Solar System. We do not know whether such events are common in planetary systems or whether this has been a unique event in the Solar System.

Traces of life have been found in rocks 3500 million years old, when the Earth had been in existence a mere 20 per cent of its life-span. Now, the chances of the amino-acids and nucleotides forming in the primeval oceans and then coming together randomly to create enzymes, the substances that cause and direct the numerous chemical reactions that occur in living organisms, has been estimated at something like $10^{78}$ against. So it appears that astronomers are hard pressed to explain just how and why life has developed at all on Earth. If it is indeed true that the chances against the evolution of life are so large, then the present optimism which leads to the search for life elsewhere in the regions of the Milky Way closest to the Sun may be quite unjustified.

However, we have only just started our search – the Universe is an unimaginably large place, so we cannot expect to find all the answers to the questions of life elsewhere so soon.

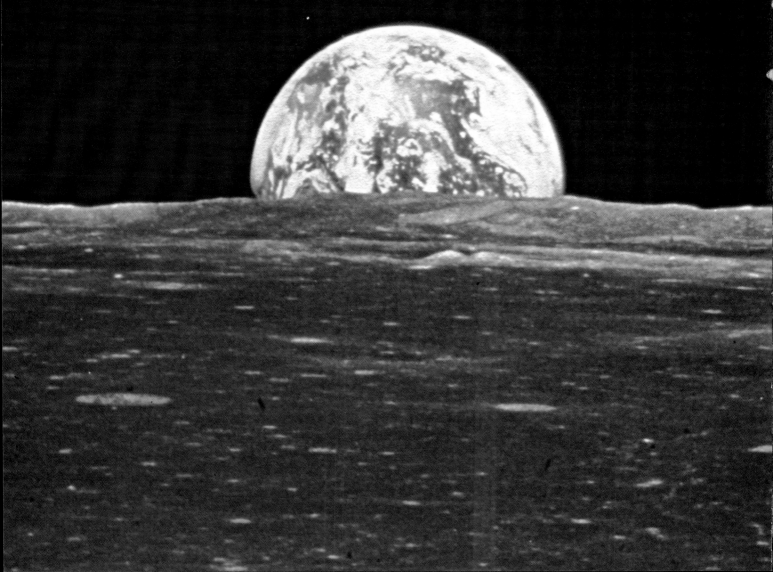

# 2/PLANET EARTH

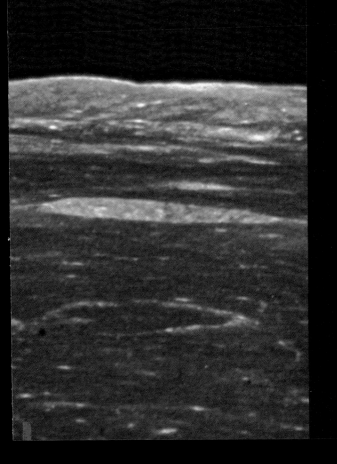

Our own planet, the third from the Sun, is a peculiar
object. It is the only one of the Solar family that lies in
the so-called 'zone of life' – the zone within which
life-giving water is (generally) neither frozen nor
evaporated, but liquid, as even a distant view from
the Moon reveals (left). It contains a wealth of
life-forms. It is reasonably equable, though the climate
does undergo long-term cycles of heat and cold. Yet it
possesses a hot interior that keeps its surface features – the
continents – in steady, if infinitesimal motion, and
provides it with a magnetic field stretching far out into
space. In these and in other ways, it is a unique planet.

Traditionally, astronomy has been the study of heavenly bodies from the surface of the Earth, and the Earth itself has not usually been considered part of the subject. Of course, the Earth is one of the planets and its annual circuit round the Sun and its daily rotation on its own axis can only be explained in astronomical terms. But beyond this, our planet has been more the concern of geologists, climatologists and biologists than of astronomers.

Since the beginning of the space-age, however, these boundaries have broken down. In recent years, scientists have come to see that there is often no clear distinction to be made between the earth sciences and astronomy. A study of the Earth's unique interior may cast light on the history of the Solar System and thus on the formation of other planets. A study of volcanoes may not only show how our own atmosphere was created, but may also give clues about the composition of the atmosphere of other planets.

We know so much about the Earth that astronomers must draw a line beyond which they can leave research to other specialists. This chapter will look at the Earth from the point of view of an outside observer – an astronomer, say, from an alien civilization, observing Earth for the first time, and seeking to explain those things about it that seem to be unique. (There are many characteristics it shares with other planets of course; these are dealt with in Chapter 4.)

Our alien astronomer would note, for instance, that the Earth is marbled with coloured areas, some of which vary with the seasons; that it is three-quarters ocean; that the continents look like pieces of a jigsaw puzzle; that the poles are ice-covered, but that the ice advances and retreats with the seasons; and that the planet possesses a gaseous blanket capable of sustaining plant life (and possibly, therefore, more complex life forms).

## Into the Earth's Interior

The interior of the Earth is totally inaccessible to direct observation. The deepest hole drilled into the surface (an oil well in Oklahoma) penetrates about 10 km. (6.25 miles), but this is only a fraction of the 6,378 km. (3,986 miles) to the centre. Information about the interior must be obtained by, in effect, X-raying it with seismic waves. These are produced by earthquakes and artificial explosions, which make the whole Earth ring like a bell. The waves have different speeds depending on the depth and density of the material they pass through.

The waves are of four types. Two of these, Love waves and Rayleigh waves, only travel near the surface and so tell us nothing about the interior. The other two types, S-waves and P-waves, can travel deep inside the Earth and can be detected at great distances from the earthquakes that produce them. A P-wave is like a sound wave in air and the vibrations are in the direction of motion, whereas they are at right angles for an S-wave. An important consequence of this differ-ence is that S-waves can only travel in solids whereas P-waves can travel in solids and liquids.

Scientists measure the waves with sensitive instruments called seismometers, and from the times at which the waves from a particular earthquake arrive at seismometers in different parts of the Earth the times for them to travel along various paths can be calculated. These times can be many minutes for the longest paths. From the travel times it is possible to calculate the speed of the waves at different depths, and from this the density of the material making up the interior of the Earth.

The most important discovery made in this way was that the density does not vary smoothly with depth but has a number of sudden jumps, showing that the Earth is made of shells having different properties. The top layer of the Earth is called the crust. In 1909 an obscure Croatian seismologist named Andrija Mohorovičić was studying earthquakes in the Balkan Peninsula when he discovered that the speed of P- and S-waves suddenly increased at a depth of a few tens of kilometres. It is now known that this is a world-wide phenomenon, although the depth varies from 30 to 40 km. (19–25 miles) under the continents and about 10 km. (6.25 miles) below the oceans. This depth marks the base of the crust and the top of the layer below, the mantle. The boundary between the crust and the mantle is now called the Mohorovičić discontinuity, or Moho for short.

Mohorovičić's methods can also be applied to study greater depths, although the network of seismic stations must be spread wider than his. In 1906 the British scientist R.D. Oldham showed from an analysis of P-wave data that the Earth has another layer, the core, below the mantle. Later, in 1914, Beno Gutenberg, a German-born geophysicist, showed that S-waves, which travel only in solids, do not penetrate into the core, although P-waves do. This means that the outer layers, at least, of the core must be liquid. Gutenberg calculated the depth of the core-mantle boundary to be 2,900 km. (1,812 miles). In 1936 Inge Lehmann, a Danish seismologist, showed that in the core, only the outer part is liquid. The inner core, from a depth of 4,980 km. (3,112 miles) to the centre, is solid.

The average density of the Earth is 5.518 times that of water, but the seismic evidence shows that the actual density varies with depth. The crust is about three times denser than water on average, while the inner core is about 13 times denser than water. The increase in density with depth is due to the enormous pressure exerted by the weight of the layers above. But the increase is not steady: there is a sudden change at the boundary between the core and the mantle which can only be explained by a similarly sudden change in composition. The outer core is believed to consist mainly of iron, probably with about six per cent of nickel, while the mantle is a mixture of minerals which are mainly compounds of the elements oxygen, silicon, magnesium, calcium

The interior of the Earth, as revealed by the way earthquake waves travel (*right*) and by other, theoretical considerations, consists of four major elements. The inner core, which is solid, and the liquid outer core both consist largely of iron. Movements in the core create the Earth's magnetic field. Outside the core lies the mantle, which consists of solid – but not rigid – rock. Its lower regions are believed to be static, but towards the surface of the Earth, the material of the mantle can flow, very slowly and over long periods of time. Convection currents in this region, the asthenosphere, circulate hot rock from below (red) to a few miles beneath the surface, where it cools (blue) and eventually sinks again. On the mantle's upper region, the lithosphere, is the crust, in which float the lighter continental plates, which are kept in slow but permanent motion by the upwelling rock beneath.

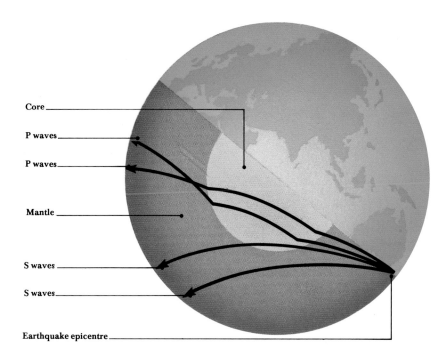

Core

P waves

P waves

Mantle

S waves

S waves

Earthquake epicentre

and iron. Again, the solid inner core is somewhat denser than the liquid outer core, but the difference in density appears to be too great to be just that between a liquid and a solid of the same chemical composition, so there must be some difference in composition also.

The temperature is about 4,000°C at the centre of the Earth and decreases steadily towards the surface. Since heat flows from high to low temperatures there is an outward flow; at the surface it is about 0.06 watts per square metre. Although this is much less than the heat received from the Sun, it is this heat from the interior that supplies most of the energy for volcanoes, earthquakes and mountain building.

It is now generally believed that the Earth formed from small solid particles with a temperature of 200°C or less. Where did the heat come from to raise it to its present temperature? One source is the energy released by the particles that

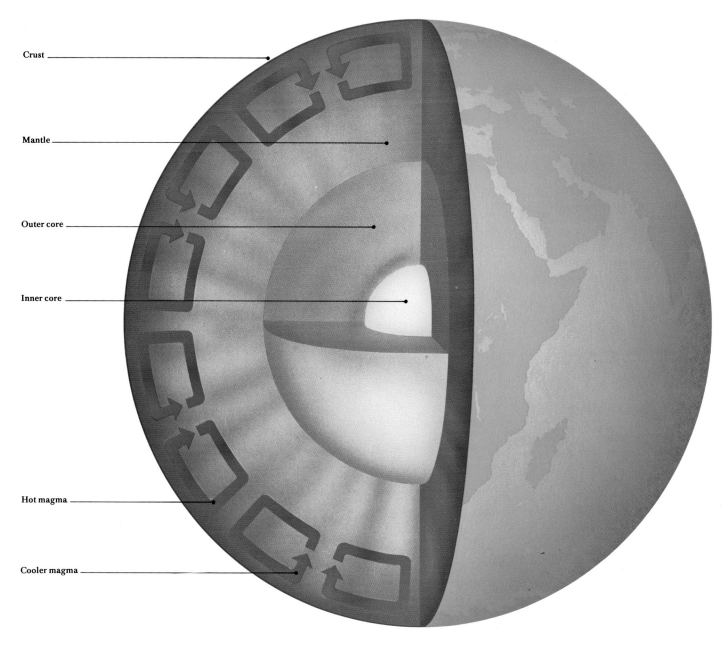

Crust

Mantle

Outer core

Inner core

Hot magma

Cooler magma

collided to form the primitive Earth. Although the total energy released in this way was sufficient to raise the Earth to a temperature of 20,000°C, it was always produced at or near the surface and was probably quickly radiated away. To heat the interior another energy source is required: radioactive decay.

Roughly speaking the radioactive elements responsible for heating the interior of the Earth are of two types: short-lived elements with half-lives of one to ten million years and long-lived elements with half-lives of 1,000 million years or so. These figures should be compared with the age of the Earth, which is 4.6 thousand million years. If there were sufficient quantities of the short-lived elements then the Earth would have quickly heated up. Since all these elements have long since decayed it is not possible to say how important their contribution was. The long-lived elements such as uranium, thorium and the weakly radioactive but abundant potassium are still present. Although the exact amounts inside the Earth are somewhat uncertain there is no question that they can provide the energy necessary to explain the present temperatures.

At some point, the release of energy by radio-active elements must have melted a large part of the Earth since this is the only way known for the separation of the original body of uniform composition into a core and a mantle. A similar process occurs when impure iron is melted in a steelworks and the non-metallic parts separate out to form a low-density slag which floats to the surface.

It was in this way that the primitive crust was formed. The commonest element is oxygen which makes up 46.5% by mass, followed by silicon with 28.9%. Other common elements are aluminium (8.3%), iron (4.8%), calcium (4.1%), potassium (2.4%), sodium (2.3%), magnesium (1.9%) and titanium (0.5%). Everything else together makes up just 0.3%. Only rarely do these elements occur as such; in general they are chemically combined into a vast range of minerals.

The Earth's internal heat has another fundamental significance: it provides power to form the crust – to shift continents, to build mountains, to cause earthquakes and volcanoes.

## The Drifting Continents

A glance at any atlas or globe is sufficient to show the apparent fit of the bulge of South America into the bight of Africa. This striking physical feature has caused many laymen and scientists to speculate that the continents were once joined together and have subsequently drifted apart. Ancient myths and legends such as the destruction of Atlantis show that in the past the idea of moving land was quite acceptable. The legends arose because people had experienced volcanic eruptions and earthquakes, and to their minds it was obvious that land moved around, that *terra* was not necessarily *firma*.

The earliest-known written suggestion of continental drift was made by the English philoso-

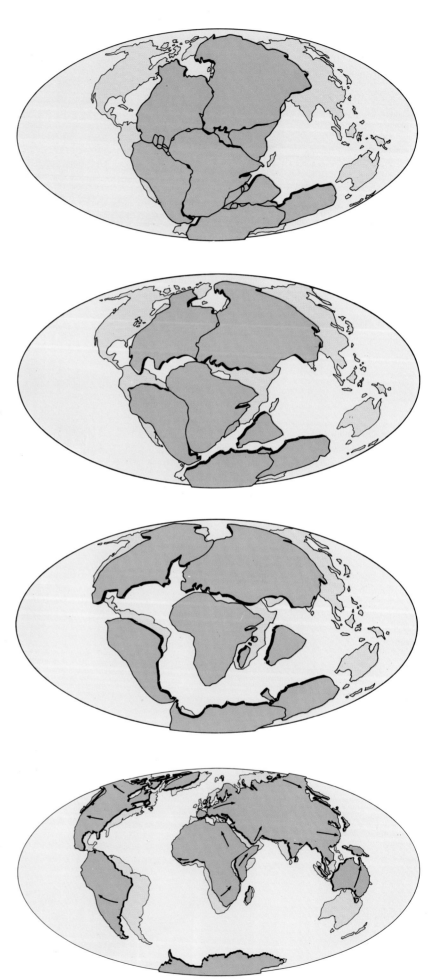

These diagrams show the movements of the continental plates over the last 200 million years, and how they will move in the relatively near future. It is now generally accepted that the plates have been in motion since they were formed over 4,000 million years ago, but the mountain remnants of previous continental collisions have long since been eroded and battered away. A solid geological baseline for 'recent' events is provided by the formation of the super-continent Pangaea ('land everywhere') which was in existence by some 300 million years ago (left).

A hundred and eighty million years ago the northern group of continents, known as Laurasia, had split away from the southern group, known as Gondwana, which itself had begun to break up. India had become an island drifting northwards, but Antarctica and Australia were still joined.

Sixty-five million years ago, North America and Europe were still linked, but the south Atlantic was well formed. Africa had swung to hit the European plate, closing an oceanic inlet to create the Mediterranean, Black and Caspian Seas. Drifting had separated Madagascar from Africa and was beginning to separate Greenland from North America.

Fifty million years hence, the Atlantic will be considerably larger and the Pacific smaller. The San Andreas Fault will have separated, isolating California, and East Africa will have split apart along the Great Rift Valley.

pher Francis Bacon in 1620, but as modern science developed during the 17th century, ancient hearsay such as the legend of Atlantis was deliberately discarded, and with it the ideas of mobile land. For nearly 300 years, the idea of continental drift was rank heresy, though it was supported by a few lone voices such as the German naturalist Alexander von Humboldt around 1800 and the Austrian geologist Eduard Suess a century later. Geology was dominated by the idea that land could rise and fall, but the notion that it could move sideways was regarded as ludicrous.

The man who really established continental drift as a serious scientific proposition was the German geophysicist Alfred Wegener with two articles that he published in 1912. These outlined his wholly original continental drift hypothesis which was by far the most detailed and comprehensive then proposed. He concluded that until 40 million years ago all the continents were joined together in one land mass, which he called Pangaea ('land everywhere'). Pangaea was like a grounded ice floe, which broke up, allowing the separate continental blocks to float away on a partially liquefied layer beneath.

Major opposition to continental drift was expressed by the English geophysicist Harold Jefferies in 1924 and his enormous prestige led to initial rejection of the theory. The response may now seem short sighted, but it had scientific validity. Wegener had suggested various forces which would cause the drifting apart of the continents, but most geologists rejected them as too small to produce the required effects. Neither he nor his supporters could come up with a theoretically sound mechanism to explain why the continents move around the globe.

Since then, the climate of opinion has undergone a revolution as profound in its way as the Copernican revolution that set the Sun at the centre of the Solar System. In the 1950's, marine geologists and geophysicists, such as Maurice Ewing and his co-workers at Columbia University, discovered that the crust under the oceans was only a few miles deep, much less than the depth typically found under the continents. They also discovered that a mid-ocean ridge, first discovered in the North Atlantic, was a feature of all the major ocean basins.

In an attempt to explain these phenomena, Professor Harry Hess of Princeton University in 1960 proposed the hypothesis of sea-floor spreading, which led to an exploration of continental drift. He postulated that the heat flow from the centre of the Earth causes convection currents in the mantle and that the crests of the mid-ocean ridges mark the positions of the rising currents. As the currents hit the surface, they spread away from the ridges, and carry the oceanic crust with them. The gap that this leaves is filled with material of the mantle to form new crust.

Hess suggested that along each ridge about 1 cm. of new crust is formed each year. At this rate, it would take a mere 200 million years to form

all the present deep ocean floor. As this time is less than five per cent of the age of the Earth, Hess concluded that old crust is destroyed at the same rate as new crust is generated.

Confirmation of Hess's hypothesis has come from observations of the magnetization of the sea floor rocks, which record the direction of the Earth's magnetic field. Now, the Earth's field is not fixed: it flips directions, for reasons not fully understood. Sea floor observations show that there are stripes roughly parallel to the mid-ocean ridge crests which are alternately magnetized in the same and the opposite direction as the Earth's present magnetic field. As mantle material rises to form new crust, it cools and is magnetized in the same direction as the Earth's field, once its temperature falls below a particular value known as the Curie point. Provided its temperature stays below this value the rock keeps its direction of magnetization, even if the Earth's field reverses. If this is true the pattern of magnetization of the two sides of a mid-ocean ridge crest should be symmetrical, a proposal made by Frederick J. Vine and Drummond Matthews in 1962 and subsequently confirmed by examination of deep-sea sediments.

It follows that if an ocean floor is growing, the continents along its boundaries must be moving apart and so the discovery of sea-floor spreading led to the general acceptance that continental drift does indeed occur.

The theory as now developed is known as the theory of plate tectonics, which differs from earlier theories in seeing the moving units as involving much more than the continental crust. The Earth's mantle has a change in its properties at a depth of about 100 km. (62 miles). The mantle and crust above this depth constitute the lithosphere, and the mantle below it the asthenosphere. The lithosphere is rigid but is broken into seven great slabs (plates) and a number of lesser ones which move about on the asthenosphere below. The movements of the plates are driven by convection currents in the asthenosphere, which, although solid, can creep at a sufficient rate for this to be possible.

As the plates move around they interact with one another along their boundaries in four main ways. One way is for two plates to move apart at a spreading ridge such as the mid-ocean ridge in the Atlantic Ocean. Here, as we have seen, new crustal material is formed and the two plates grow along their common boundary. At other boundaries, two plates collide. If one of them contains oceanic material it is thought that it is forced down below the other in a subduction zone. At the boundary itself there is a trench along the ocean floor. The oceanic plate can be carried down as far as 700 km. (437 miles) before it completely breaks up. The crustal material that is taken down is partially melted and, being less dense than the surrounding mantle, it then rises again towards the surface. Much of it erupts as lava and builds up a chain of volcanic islands, for example the Aleutian, Kurile, Japanese and

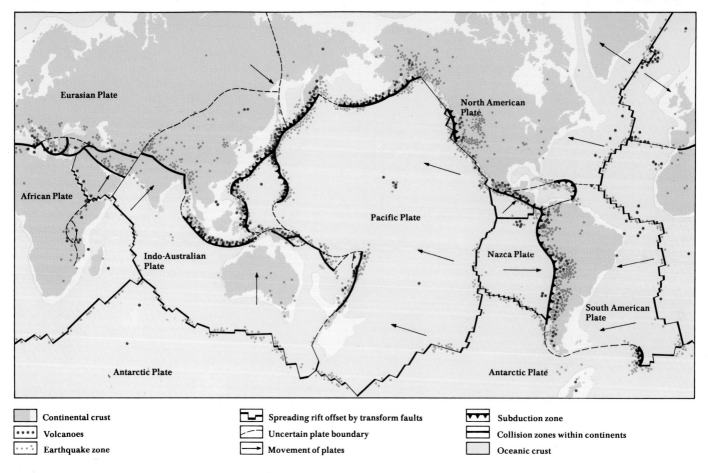

| | | | |
|---|---|---|---|
| ▢ | Continental crust | ⊏▔⊐ Spreading rift offset by transform faults | ▽▽▽ Subduction zone |
| •••• | Volcanoes | ⟋⌁ Uncertain plate boundary | ▭ Collision zones within continents |
| ⋯⋯ | Earthquake zone | → Movement of plates | ▢ Oceanic crust |

The ten major plates that make up the Earth's crust are clearly defined by volcanoes and earthquakes. The plates are in continuous motion, jostled by the motion of the hot, semi-fluid rock beneath. Above upwelling currents, plates move apart, the gaps being filled by hot rock from below. Along other edges – like that marked by the so-called 'ring of fire' circling the north Pacific – the rocks of the lithosphere plunge back into the depths of the Earth, leaving the bulk of the continental rocks unaffected.

Marianas islands. However if the subduction zone directly borders a continent, as along the Pacific coast of South America, the volcanoes form on land, in this case along the Andes mountain range. Friction between the plates at subduction zones causes intense earthquake activity (as in the western parts of South America).

In those areas where two continental plates collide, neither is pushed under the other. Instead, there is a collision zone where the plates crumple and mountain ranges, such as the Himalayas, are formed. Since oceanic crust is being destroyed at a subduction zone the whole ocean is eventually consumed and the zone is converted into a collision zone. Eventually, the Pacific will vanish.

The fourth main type of plate boundary is a transform fault where two plates slide alongside one another with no creation or destruction of plate material. These faults typically occur at intervals along a spreading ridge and provide a series of offsets. A notable example is the San Andreas fault in California. All types of plate boundary are notable for being active zones and almost all the earthquake, volcanic and mountain-building activity of the Earth's crust closely follows them. Almost as a by-product of this activity the continents are carried round as passengers on the plates so that continental drift is a normal part of the Earth's behaviour.

Plates can break and split apart with the formation of a new ocean between them. This has happened in the past; for example South America and Africa split apart when the South Atlantic Ocean began to open about 135 million years ago. The most recent split is thought to have occurred about 40 million years ago when

Australia and Antarctica separated. 200 million years ago the Atlantic, Indian, Arctic and Antarctic Oceans did not exist. The Pacific however is much older and has been closing as the other oceans have opened.

In contrast to the relative youth of much of the ocean floor, most of the continents are much older. A surprisingly large proportion of the continental crust was in existence 2,500 million years ago. Plate movements have since then welded on successively younger extensions.

Predictions of future movements are also possible. In about 10 million years, motion along the San Andreas fault will bring Los Angeles alongside San Francisco. Some geologists have predicted that as Africa rotates anticlockwise, the Red Sea and the Gulf of Aden will widen while the western Mediterranean will swing shut, closing off the Strait of Gibraltar. Finally, the African Rift Valleys will open.

This theory not only neatly explains why the continents are as they are; it also explains why volcanoes occur, which are, as it turns out, vital to the evolution of the Earth in their own right: they are in part responsible for the constitution of the atmosphere.

**The Changing Atmosphere**
Near the Earth's surface the atmosphere consists of nitrogen (78.1% by volume), oxygen (21.0%), argon (0.9%) and a small amount of carbon dioxide (0.03%). There are also small variable amounts of water vapour and trace quantities of methane, nitrous oxide, carbon monoxide, hydrogen, ozone, helium, neon, krypton and xenon. The pressure falls off with height and at 6 km. (3.75 miles) it has only half its sea-

# Eclipses: Shadows of the Earth and Moon

The Earth and the Moon throw long conical shadows into space away from the Sun; when one of these bodies moves into the shadow of the other there is an eclipse. Each shadow has two parts, an *umbra* in which the Sun is totally obscured and a *penumbra* where the obscuration is only partial.

As the Moon goes round the Earth it usually passes above or below the Earth's shadow, but sometimes it goes right through it. There is then an eclipse of the Moon. Such an eclipse starts when the Moon enters the Earth's penumbra. Some of the light from the Sun is cut off so that the Moon becomes darker, but the effect is not generally enough for an observer on the Earth to notice unless he has been forwarned. Later the Moon enters the umbra. There is a partial phase to the eclipse while only part of the Moon is in the umbra but this is followed by a total phase with the Moon completely immersed. No light from the Sun can directly reach the Moon when it is in the umbra but some is scattered by the Earth's atmosphere into the shadow. Different colours are scattered by different amounts and as a result the totally eclipsed Moon is not completely dark but has a distinctive coppery-red hue. The total phase is followed by second partial and penumbral phases as the Moon leaves the Earth's shadow. The total phase can last as long as 1 hour 40 minutes.

Viewed from the Earth, the Moon generally passes above or below the Sun, but it sometimes passes in front and casts its shadow on the Earth. Purely by chance, the Moon and the Sun appear to have very nearly the same size when seen from the Earth and under favourable circumstances the Moon completely covers the Sun and there is a total eclipse. As the Moon moves in front of the Sun its shadow sweeps across the Earth in a narrow path, never more than a few hundred kilometres across.

An observer on the Earth sees a gradually increasing bite taken out of the Sun. As totality is reached, the visible part of the Sun is reduced to a thin crescent. Immediately before totality this crescent is broken by the irregularities of the mountains at the edge of the Moon's visible disc to form what are known as Baily's Beads (after the early 19th-century English astronomer, Francis Baily). The last visible part of the Sun often gives the appearance of a diamond ring. During totality the Sun's atmosphere, the corona, which is normally masked by the glare, becomes visible in all its splendour. After totality the partial phases are repeated in reverse order. At the equator the total phase can last as long as 7 minutes 40 seconds, but this maximum time becomes smaller towards the poles; for example, at latitude 45° it is 6 minutes 30 seconds.

Since their orbits are not completely circular, the distances of the Moon and the Sun from the Earth vary somewhat. Their relative sizes also change. At a total eclipse of the Sun, the Moon only occasionally appears slightly larger. More often, the Sun is visible all round the Moon at mid-eclipse. This is known as an *annular* eclipse.

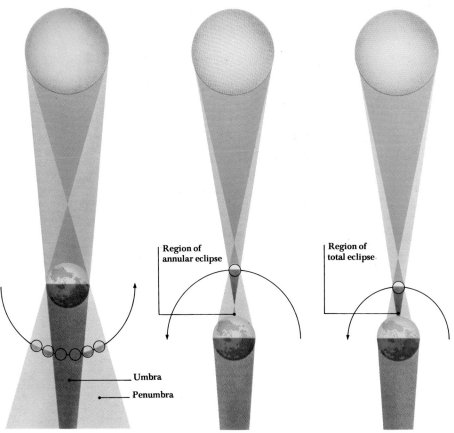

**Region of annular eclipse**

**Region of total eclipse**

Umbra
Penumbra

During a total eclipse, as this sequence shot in 1973 shows, the Moon exactly covers the Sun's disc.

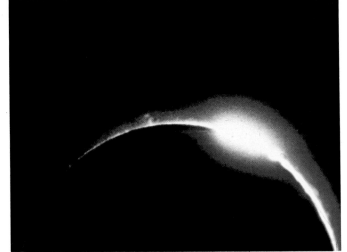

The Sun shining between lunar mountains creates a Baily's Bead.

At totality, the Sun's corona (upper atmosphere) becomes visible.

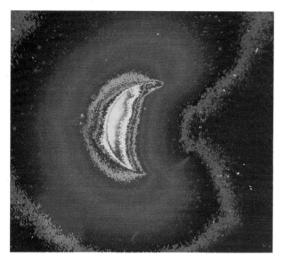

The Earth's atmosphere, which is constructed in four zones and several sub-zones, acts as a blanket that preserves the surface of the Earth from the effects of the most destructive forms of radiation. Atmospheric molecules of hydrogen have been found up to 5,000 miles from Earth, as the ultraviolet picture above, shot from Apollo 16, shows (the different layers of colour reveal temperature differences). But most of the atmosphere, as the diagram at right shows, is held near the surface of the Earth by gravity – 75 per cent is below the height of Everest.

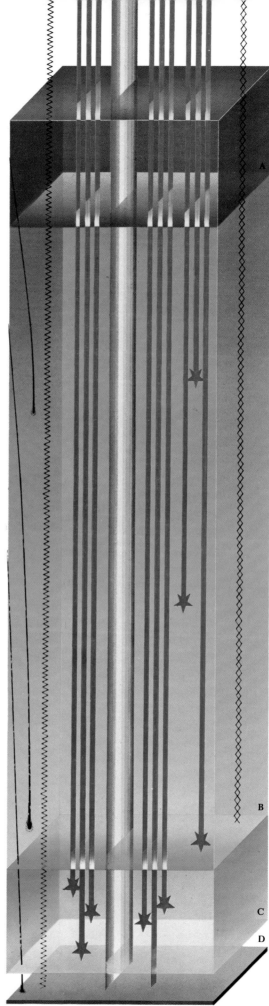

**A.** The exosphere extends upwards from about 640 km. (400 miles) merging with the interplanetary medium. It consists largely of diffuse hydrogen and helium.

**B.** The ionosphere, which spans the region from 80 km. (50 miles) to 640 km. (400 miles), is an electrically conducting region. Here the atoms and molecules have had their electrons removed by solar radiation.

**C.** In the stratosphere, ultraviolet radiation from the Sun converts oxygen into ozone, thus absorbing harmful radiation. This reaction raises the temperature of the stratosphere to about 0°C., acting as a 'lid' on the weather systems of the troposphere below.

**D.** The troposphere extends from the surface up to about 16 km. (10 miles). This unstable region, subject as it is to the differences in temperature and moisture content of the Earth's surface, is responsible for almost all the Earth's weather. Only radio waves, visible light and a few ultraviolet rays penetrate to this level.

level value. Although the pressure and density drop to very small values at greater heights the atmosphere is still detectable several thousand kilometres above the surface by, for example, the drag it exerts on artificial satellites. There is also a change in composition with height. Below 500 km. (312 miles) oxygen and nitrogen are still the most common elements. Further up helium and then hydrogen are the main constituents.

When we try to discover where the Earth's atmosphere came from and how it has evolved to its present composition we find little direct evidence. It seems unlikely that the Earth captured an atmosphere when it was formed; if it had, then the proportions of heavy gases, such as neon, argon and krypton would have been much greater than they are.

The alternative is that gases were trapped in the rocks and later released. As we have already discussed the Earth must at some time in its past have been almost, if not wholly, melted in order that differentiation into a crust and mantle could occur. The same melting would also cause the release of gases trapped in the rocks to form an atmosphere. This outgassing (as it is called) is also caused by volcanoes and meteoritic impacts. It seems that most of the surface water was released in this way. There is a difference of opinion as to whether carbon dioxide and nitrogen were released in the same way or whether they were produced from methane and ammonia.

Perhaps the most puzzling problem raised by the constituents of the atmosphere is that of oxygen. Oxygen is vital to all forms of animal life; yet it was not part of the primitive atmosphere. Although oxygen is the major constituent of the outer layers of the Earth it was totally locked up in other compounds. Because free oxygen is a very reactive chemical it would have left easily detectable effects, which we do not in fact see. For example, the primary organic molecules necessary for the origin of life would have been destroyed by oxygen, and many old rocks consist of compounds that would have been oxidized if there had been any oxygen present when they were formed. We know therefore, that until about 1,800 million years ago, long after evolution of the first simple life-forms, there was no free oxygen in the atmosphere.

A number of processes (e.g. photosynthesis in green plants) can split up carbon dioxide or water and release oxygen. However, unless some of the products of these reactions are removed they recombine with the oxygen and there is no overall change in atmospheric composition. Only in the case of photosynthesis is it thought that

this occurs. Some of the plant material has been converted into limestone, coal, oil and other minerals leaving the oxygen in the atmosphere. It is a remarkable thought that the Earth was not a ready-made home for creatures like us. It had to be modified by earlier forms of life before it was suitable.

**A Magnet in Space**
Another peculiarity of the Earth which is also caused by its internal structure is its magnetic field. The Earth's magnetic field is similar to that produced by a bar magnet or a uniformly magnetized sphere, a fact first pointed out by William Gilbert, who was physician to Queen Elizabeth I. It has two magnetic poles which are close to, but not exactly at the Earth's geographic

# The Rhythm of the Tides and Seasons

The tides (right) and the seasons (below) are together the two most obvious consequences of the interaction of the Earth, Moon and Sun.

Although the Moon is so much smaller than the Sun, its proximity to the Earth gives it twice the pull of the Sun on the Earth's waters. The Moon's gravity pulls the oceans into two bulges, one on the side facing the Moon, the other on the opposite side. The Sun's influence is generally ironed out by the greater influence of the Moon. But twice a month, at new and full Moon, the Sun and Moon pull in line and their tidal effects produce tides that are higher than normal (spring tides). When the Moon is at first and last quarter, it pulls at right-angles to the Sun and the tidal effects work against each other producing a small range (neap tides). The Earth rotates beneath the bulges, creating the familiar sequence of high and low tides.

The rhythm of the seasons has nothing to do with any change in the distance from the

Earth to the Sun. Although the Earth's orbit is not a precise circle, the deviation – and thus the change in heat received – is infinitesimal. The true cause of the seasonal change is the tilt of the spinning Earth, about $23\frac{1}{2}°$ away from the perpendicular. Consequently, first one hemisphere and then the other leans towards the Sun.

As the 'summer' hemisphere, for instance, begins to tilt inwards, the Sun, seen from the Earth, climbs higher in the sky. This firstly makes the proportion of daylight hours a greater fraction of the whole 24 hours, reaching an extreme inside the Arctic and Antarctic circles, where the summer Sun never sets. Secondly, it brings the Sun's rays down more nearly vertical at local noon, concentrating their heating effect rather than spreading the warmth over a slanting path.

In the winter, the opposite happens: daylight hours are reduced (the poles lie in permanent darkness for several months), the Sun only rises low above the horizon, and its heat is weakly dissipated over a slanting path.

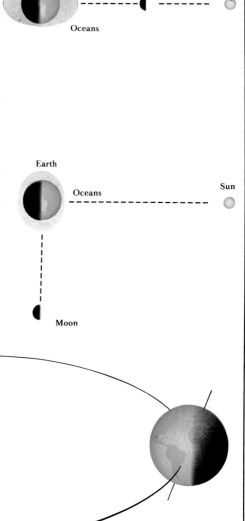

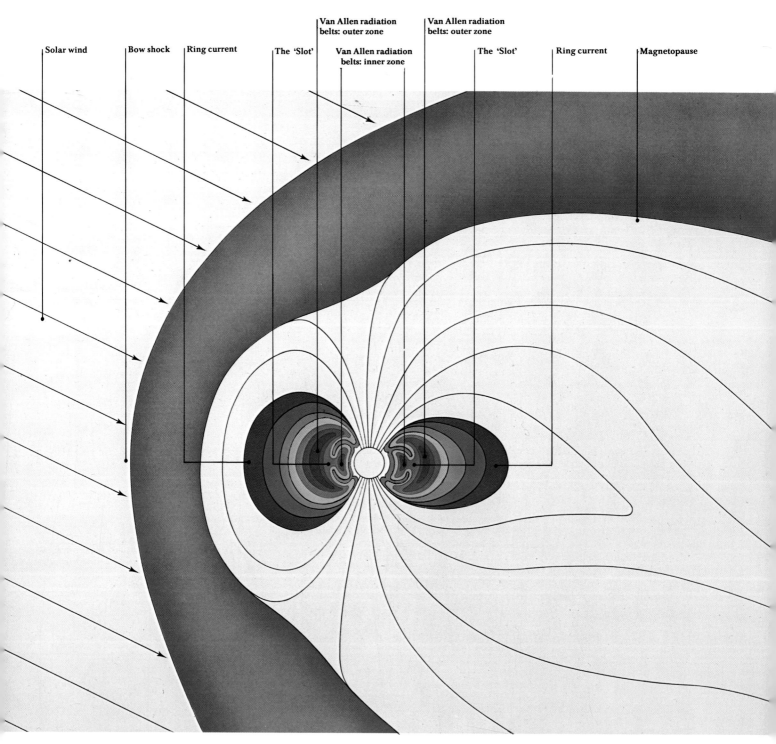

Solar wind | Bow shock | Ring current | The 'Slot' | Van Allen radiation belts: outer zone | Van Allen radiation belts: inner zone | Van Allen radiation belts: outer zone | The 'Slot' | Ring current | Magnetopause

poles; for example the north magnetic pole is at present in Canada 1,300 km. (812 miles) from its geographic counterpart. Scientists have been measuring the strength and direction of the magnetic field for about 400 years, during which time there have also been quite substantial changes in the position of the magnetic poles. The rapidity of these changes shows that their cause cannot be in the solid outer layers of the Earth. This is further confirmed by the irregularities in the field which are quite unconnected with surface features such as continents and oceans.

Some rocks are naturally magnetic (some of them, as we have seen, lie on the ocean floor) and when they were formed they were magnetized

in the direction of the magnetic field at the time. This magnetization has been maintained to the present day by many rocks. Paleomagnetism, which is the study of this fossil magnetism, can give scientists information about the magnetic field millions of years ago. One of the most amazing outcomes of such studies was the discovery that about once every 400,000 years the field completely reverses in direction, a phenomenon first observed by the Frenchman Bernard Brunhes in 1906. During the last 71 million years there have been 171 such field reversals, the most recent of which was 700,000 years ago. Why the reversals occur is at present a mystery.

The Earth's influence spreads far beyond the confines of its own surface. Besides its gravitational field, it also possesses a complex magnetic field, the magnetosphere, which begins some 300 miles (500 km.) above the surface overlapping the upper regions of the atmosphere. As in a bar-magnet, lines of force arc out from the north and south magnetic poles, creating a region that can be detected by special photographic film (*left*). But the field is distorted. In its orbit, the Earth is swept by radiation from the 'solar wind', which flattens the leading edge of the field in to about 70,000 km. (40,000 miles) from the Earth, while away from the Sun the field forms a tail hundreds of thousands of miles long. The field deflects most of the solar wind but some particles enter the Earth's atmosphere at the poles causing dramatic electrical displays, aurorae (*right*). Some particularly energetic particles penetrate the magnetic shield and become trapped in two intense zones, the van Allen radiation belts.

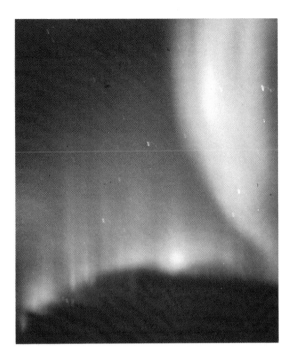

As we have seen, the origin of the magnetic field is not to be found near the surface of the Earth. Because of the continuous changes and frequent reversals of the field the obvious place to look for its origin is in the liquid outer core where we can hope for motions that are much more rapid than those that can occur anywhere else.

The first problem is to find out the mechanism that maintains the field. Permanent magnetism is not believed to be possible in a liquid, and even if it were the different parts would soon be mixed up and there would be no general magnetism. The other way of producing a magnetic field is by electric currents. The Earth's core is mainly iron and nickel, so being metallic, it is a good conductor of electricity. However, calculations show that if an electric current were started in the Earth's core it would die away after several tens of thousands of years. This is such a small fraction of the age of the Earth, which is 4,600 million years, that there must be some force driving the currents.

The general view today is that the field is caused by what scientists call a self-exciting dynamo. In a man-made dynamo, when an electrical conductor is driven round in a magnetic field it generates an electric current. But part of the electric current can in its turn be used to generate the magnetic field, which then creates more electricity. (This is what happens in the generators at a power station.) It is possible to connect two dynamos in such a way that, with the same direction of rotation, the electric currents and magnetic fields can flow in either direction and can even spontaneously reverse from time to time. Such a system has a complex structure, which is not true of the Earth's core; but there could be complex motions of fluid in the core that compensate for the simplicity of structure to produce the field as we measure it. Scientists are

now trying to discover if there are any such motions in the core.

Whatever its cause, the Earth's magnetic field extends well above the surface. The Sun continuously emits a stream of charged particles known as the solar wind. As the stream flows past the Earth, the magnetic field – like a big, soft ball in a current of water – is flattened on its leading side and stretched on the far side. The region within which this occurs is called the magnetosphere. The boundary of the magneto-sphere, the magnetopause, is about 70,000 km. (43,750 miles) from the Earth on the sunward side, but on the other side the magnetosphere's tail extends beyond the Moon's orbit.

The magnetic field strongly affects the paths of the charged particles, such as cosmic rays, that approach the Earth. Those of low energy, such as most of the protons in the solar wind, cannot enter the Earth's field at all. More energetic particles can enter but more easily near the poles than at the equator. As they enter the upper atmosphere, they produce an electrical display as if in a giant neon light – curtains of colour known as aurorae. These Northern and Southern Lights (Aurora Borealis and Aurora Australis) are common in polar regions, and sometimes seen in temperate zones as well.

The paths of the particles are generally very tortuous but there are certain regions where particles can follow fairly stable orbits. These regions have crescent-shaped cross-sections. Particles are trapped in them and they are called Radiation Belts or Van Allen Belts. Evidence for the existence of these belts came from the earliest Russian and American satellites, and in 1958 the American Dr. J.A. Van Allen and his colleagues provided the first picture of their shape and distribution.

### The Rhythm of the Ice Ages

If our alien astronomer observed the Earth for a very long time – long enough to measure the continents moving – he would also notice that the polar ice would sometimes spread over a much larger area, and that glaciers would flow down from high mountains to blanket the surrounding countryside in ice. He could, in other words, notice the recurrence of ice ages, the last of which ended some 10,000 years ago.

The causes of ice ages have not been identified with certainty. But *within* major ice ages the ice ebbs and flows – we are, for instance, at present in an 'interglacial' period within a major ice age, and the ice may return to northern latitudes in another few thousand years. The ebb and flow of ice probably owes a good deal to the peculiarities of the Earth's spin and orbit, peculiarities which affect the amount of heat we receive from the Sun.

This idea appeared in its classic form in the work of Yugoslav astronomer Milutin Milanko-vich in the 1930's, but became established in the eyes of professional meteorologists only in the 1970's. The theory has been summed up by

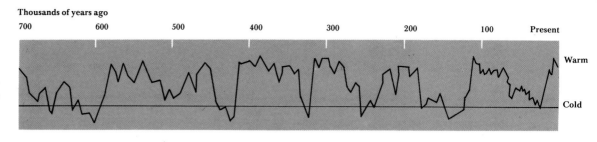

Thousands of years ago

700      600      500      400      300      200      100      Present

Warm

Cold

Though the causes of ice ages remain obscure, one theory currently in favour – the so-called Milankovich model relates the minor recurrence of ice ages to small irregularities in the Earth's spin and orbit. Every 26,000 years, the Earth wobbles like a spinning top; every 40,000 years, its tilt varies slightly; and about every 100,000 years its orbit becomes more elliptical. These effects combine to reduce the amount of heat the Earth receives from the Sun, triggering a slow but inexorable spread of polar ice.

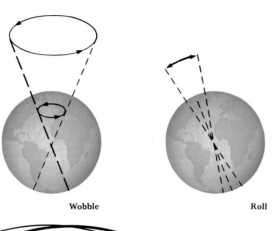

Wobble      Roll

Stretch

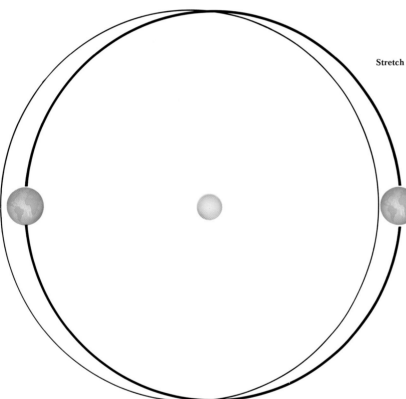

Professor B.J. Mason, the Director-General of the UK Meteorological Office.

The Milankovich Model, as it is still called, explains detailed changes in ice cover *within* a full ice age in terms of changes in the orbit of the Earth, and the orientation of our planet relative to the Sun. Three separate cyclic changes in the Earth's movements combine to produce the overall variations in the solar radiation falling on the Earth which are the key to the theory.

The longest is a cycle of between 90,000 and 100,000 years over which the orbit of the Earth around the Sun 'stretches' from more circular to

more elliptical and back; next, there is a cycle of some 40,000 years over which the tilt of the Earth's spin axis – the cause of the seasons – varies as the Earth 'nods' up and down relative to the Sun; finally, the combined pull of the Sun and Moon on the Earth causes our Planet to wobble like a spinning top as it orbits around the Sun, with a rhythm 26,000 years long. These effects combine to produce changes in the amount of heat arriving at different latitudes in different seasons.

Professor Mason has calculated just how much 'extra' heat is needed to melt ice at high latitudes when an interglacial begins, and just how much summer heat must be 'lost' to account for the advance of the ice nearly 100,000 years ago. A variety of pieces of evidence, such as the record of changing sea levels and the scratches left by glaciers in the rocks, tell geologists how much ice was around in each millennium over the past 100,000 years, and the Milankovich Model tells how much heat was arriving from the Sun each season. When Mason put the two sets of figures together, the agreement was impressive.

Between 83,000 and 18,000 years ago, the 'deficiency' of northern summer heat (insolation) added up to a staggering $4.5 \times 10^{25}$ calories (45 followed by 24 zeros); dividing this vast number by the equally vast weight of the ice sheets which built up in the northern hemisphere at that time, Mason arrived at a more every-day figure; about 1,000 calories for every gram of ice formed. And, when a gram of water vapour is cooled all the way down to a gram of ice, the amount of heat involved in the change is 677 calories. The Milankovich effect is more than enough (but only *just* more than enough) to account for the great freeze-up from 83,000 to 18,000 years ago.

What happens when the ice melts? Much less 'extra' summer insolation is needed, because the ice only has to melt to water and run off into the sea, it doesn't have to be evaporated all the way back into vapour from its frozen state. From 18,000 years ago to the present, $4.2 \times 10^{24}$ 'extra' calories were available from the changes of the Milankovich cycles; and the amount of heat needed to melt the ice that geologists know was melting over that time turns out to be $3.2 \times 10^{24}$ calories. Again, the Milankovich effect is just a little bit bigger than the minimum needed to do the job.

The model also gives us a long-range forecast – the present interglacial is just about over and we are heading back into a period of colder northern and southern winters, with a new full ice age looming up a few thousand years ahead.

# SPACE AGE VIEWS OF CLOUD AND LAND

Satellite photography has already transformed major Earth sciences. Even without the benefits of work done previously on Earth, views of the Earth from space give a vivid impression of conditions below. Scientists of an alien civilization, reporting on the Third Planet, would immediately conclude they had found a dynamic world whose wealth of vegetation and life-giving water would make it ideal for higher life forms. They might even comment (as Earth scientists often do when confronted by pictures like these) that the new-found world, with its combination of soft, marbled colours and jigsaw puzzle shapes, was one of startling beauty.

Alien geologists would soon guess that continents now separated by hundreds of miles of ocean were at one time joined. Meteorologists recording the variety of cloud forms – from swirling hurricanes to fluffy cumulus – could picture the weather systems, temperatures and wind-speeds of the atmosphere. Biologists would point out that clouds bear water from ocean to land, creating an efficient hydrological cycle. Ground cover, from lush green jungles to arid, brown deserts are easily visible.

By day, no signs of higher life forms can be seen. But an orbiting telescope – or a night time view of the glow of major cities – would soon provide evidence that at least one species on the world beneath had already evolved a technological capacity.

Above a scattering of cumulus and nimbus clouds, Skylab 1, solar panels extended, hangs in orbit.

*Above:* Moonrise from Earth orbit produces an image distorted by the lens of the Earth's atmosphere.

*Above right:* At sunset, the upper atmosphere scatters the Sun's blue light more than the red, producing a band of cerulean blue familiar from the surface.

*Right:* An extraordinary view taken just after sunset reveals turbulence in upper-atmosphere clouds.

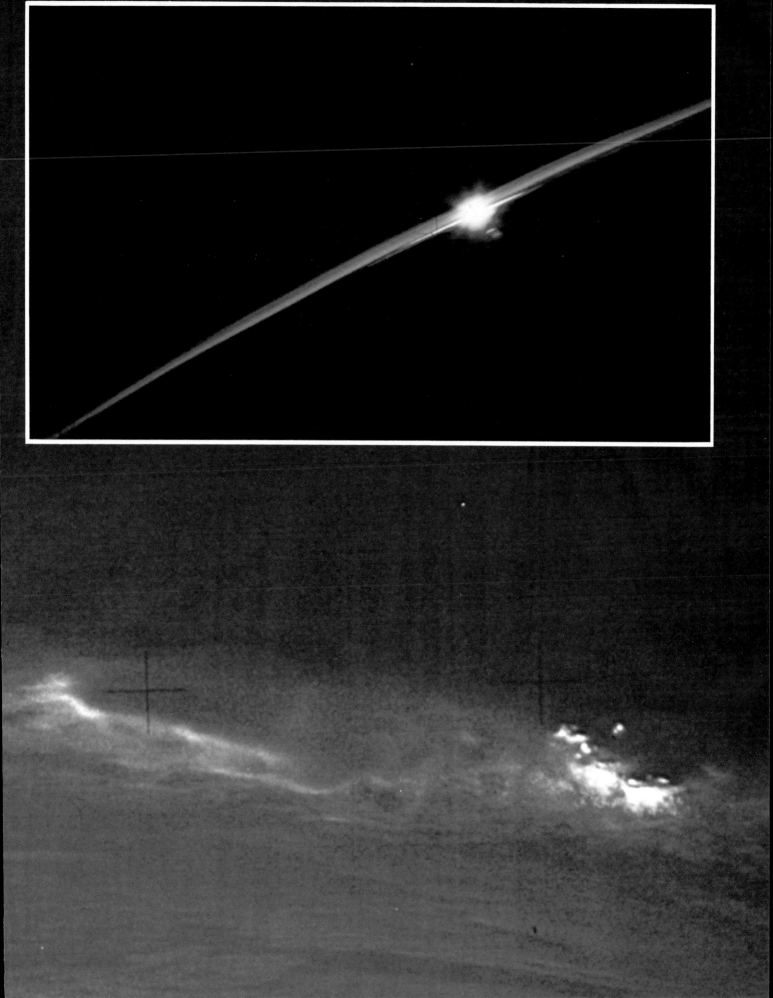

*Left:* A panorama looking south down the Nile, with the Red Sea on the left, shows the green of well irrigated areas contrasting with the desert aridity to the west (*right*) and south (*top*).

*Below:* Even at a distance of several thousand miles, the sandy wastes of the Sahara and Saudi Arabia are clearly visible.

# 3/THE SOLAR SYSTEM

Our Solar System is a collection of planets, comets and
assorted chunks of rock orbiting the Sun, all in much the
same plane. It seems likely that most of these bodies – larger
versions of the chunks of rock and dust that form the rings of
Jupiter, Saturn (left), and Uranus – are the debris left over
from the formation of the Sun, more than 5,000 million
years ago. Despite the similarity of their origin, however,
there are startling differences between the various bodies in
the Solar System, and in particular between the individual
planets.

It is something of a paradox that 20th-century mortals have less direct experience of the members of the Solar System than did our distant ancestors. The 20th-century man in the street lives in large cities, where the night sky is swamped by the glare of street lights, and the air is thick with pollutants. His ancestors had the advantages of undefiled skies, without glare or fumes, and could see the heavens with a clarity that is now almost unattainable. Apart from this, 20th-century man is also a long way divorced intellectually from the Solar System: it is no longer important for him to know what a point of light in the night sky is, or how it moves, or where it will be in a month's time. In the past, such factors were of critical importance, and were used astrologically both in constructing calendars and in the planning of future actions.

## The visible Solar System

What, then could one of those ancient observers have seen of the Solar System? What can *we* see, if we choose to make the effort and spend some hours on crisp, clear nights scanning the skies?

First, and most obvious, there is the Moon. One could hardly miss that, and its presence is taken for granted by most people. The ancients, however, were seriously puzzled by the Moon, because of its apparent changes of shape from a thin crescent to a full, glowing disc, and different tribes and civilizations devised a myriad of myths and legends to account for these, and for the curious mottled markings on its surface. Men have now been to the Moon, so much of its mystery has been lost, but there are still a great many misconceptions about it amongst people who should know better: there are some who still think that the Moon only 'comes out' at night, an error an ancient observer would never have made.

Next in importance is the great vault of the heavens, the starry firmament. The stars themselves do not concern us here, except in so far as five of them were singled out as anomalous by even the earliest observers of which we have record. The "stars" changed their positions each night. These were the five "wandering" stars, or as we now know, planets. (The word *planet* comes from the

The planets and their orbits drawn to scale reveal the pattern of the Solar System. The inner or terrestrial planets are all close to the Sun and small, the lighter gases probably driven off as they formed. The outer planets are all gaseous giants, except for Pluto. Comparable in size to our Moon, Pluto's path is eccentric – for some 20 years in each orbit it is closer to the Earth than Neptune (e.g. January 1979–March 1999). The Sun is ten times the diameter of Jupiter (*below*).

Mercury

Venus

Earth

Mars

Jupiter

| Mercury | Venus | Earth | Mars | Jupiter |
|---|---|---|---|---|
| 58/36 | 108/67 | 150/93 | 228/142 | 778/486 |
| 0.24 | 0.61 | 1.0 | 1.88 | 11.86 |
| 59 days | 243 days | 23.9 hrs. | 24.6 hrs. | 9.8 hrs. |
| 4,880/3,050 | 12,040/7,600 | 12,750/7,900 | 6,787/4,242 | 143,000/89,000 |
| 0.05 | 0.81 | 1.0 | 0.11 | 317.8 |

original Greek word for *wanderer*).

Of the five planets, Venus is much the most conspicuous. It often hangs as a brilliant jewel in the evening or morning sky, shining so brightly that it may cast its own shadow, and is regularly mis-identified as a UFO. Venus is also conspicuous because it wanders across the sky rapidly relative to the starry background and to the Sun. On one occasion, it may blaze brightly high in the sky long after the Sun has set; over the course of months it will appear to move nearer and nearer the Sun, and will set sooner and sooner after it. Eventually, it will become all but invisible, but will reappear shortly after as a *morning star*, rising before the Sun. This strange behaviour confused many early observers, not surprisingly, and many of them thought there were *two* separate bodies involved – the astronomers of ancient Greece retained two names for Venus for a long time, calling it Hesperus as an evening star, and Phosphorus as a morning star.

Next most obvious of the planets is Jupiter, shining like a particularly brilliant yellow star. Its movements, however, are relatively regular –

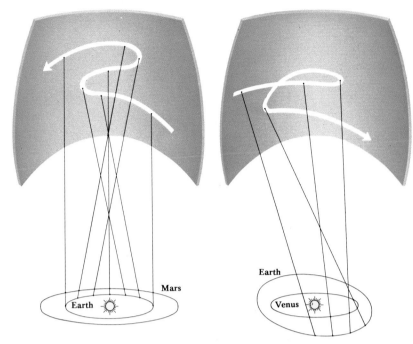

**Motion of Mars**  **Motion of Venus**

These diagrams show why a planet's motion seems peculiar when seen from Earth. As the Earth overtakes – or is overtaken by – another planet, an observer will see the planet reverse its direction against the background of fixed stars. It is not surprising the Greeks called our solar companions 'planetes', or wanderers.

| **Saturn** | **Uranus** | **Neptune** | **Pluto** | |
|---|---|---|---|---|
| 1,427/892 | 2,870/1,800 | 4,496/2,800 | 5,900/3,700 | Millions km./miles from Sun |
| 29.45 | 84.01 | 164.79 | 247.7 | Year (Earth years) |
| 10.2 hrs. | 10.8 hrs. | 15 hrs. | 6.4 days | Rotation period (Earth units) |
| 120,000/75,000 | 52,000/32,000 | 50,000/31,250 | 6,000/3,750 | Diameter (km./miles) |
| 95.1 | 14.5 | 17.2 | 0.11? | Mass (Earth = 1) |

Comets – like this one seen over France in 1811 – have always been matters of note, and usually of astrological significance, partly because they were so little understood. Until the 16th century, they were considered to be atmospheric phenomena.

The Moon in its double orbit – round the Earth and round the Sun – describes a curved path through space and shows a series of light-and-shadow phases to an Earth-based observer.

it drifts slowly against the starry background, shifting almost imperceptibly from night to night, but steadily. Mars is much more erratic. Sometimes it blazes out much more strongly than Jupiter, with the fiery red colour that has led Mars always to be associated with blood and war. It also moves highly erratically, sometimes moving slowly and steadily in one direction, then slowing down, appearing to come to a complete stop before moving backwards for a short way, and then resuming its original steady drift against the starry background.

Saturn is much less bright than either Mars or Jupiter, and looks like an ordinary star. Its planetary nature, however, is soon revealed by its movement, which is slow and regular. Much the most difficult planet to observe is Mercury. It is always very close to the Sun, and is only visible shortly before sunrise, and shortly after sunset. Usually, it is lost against the glare of the Sun, and this has caused some astronomers, perhaps plagued with more than usually bad weather, never to have observed Mercury in their lifetime. With ordinary luck, however, Mercury can often by picked out as a tiny pink lamp gleaming against the glow of the morning or evening sky. Like Venus, its appearance as either a morning or evening star confused the ancients into thinking that two bodies were present. As for Venus, the Greeks had two names for Mercury: Hermes and Apollo.

To the first observers, then, the Solar System consisted of the Moon and five points of light, moving across the heavens. Notice that we have said nothing about sizes or distances or masses or densities, or even about why the planets appear to move as they do: they are just point of light. But what else can be seen? What else is there in the Solar System?

For the naked eye observer, there are only two other visible kinds of objects: meteors and comets. On any clear, crisp night a patient watcher will see several brief flashes of light, or shooting stars, sweeping across the sky, vanishing even as they are perceived. More rarely, a great fireball may be seen, lighting up the whole sky, and more rarely still ear-splitting detonations may be heard, and solid fragments (meteorites) may reach the ground. It was not until the 19th century, however, that any link was established between these exceptionally rare 'stones from the sky' and shooting stars, which are common-place.

Rarest of all objects in the night sky are comets, which appear as diffuse points of light trailing great banners or tails across the sky. The greatest comets are perhaps the grandest of all natural spectacles, and throughout history they have been associated with profound events affecting man. It may even have been a comet that the Three Wise Men saw when they 'followed the Star' to the birthplace of Christ. More prosaically, the word 'comet' comes from a Greek root meaning something like 'long-haired star', an apt description, but not one that gives much of a clue to the nature of comets.

The visible Solar System, then, consists of planets, meteors and comets, all of them points of light moving against the great canopy of the night sky. But what is the relationship between these very disparate bodies?

We know now that the planets all move around the Sun, in regular, nearly circular orbits. It may even seem obvious that they do. But it has not always been so. For most of recorded history, men have assumed that the Earth is the centre of the Solar System. About 300 BC, one of the greatest of Greek astronomers, Aristarchus of Samos, realized that the Sun, not the Earth, lay at the centre of the Solar System, and that all the planets revolved around the Sun. Sadly for science, his arguments were not accepted, and for nearly the next 2,000 years men continued to believe what seemed to them to be obvious and

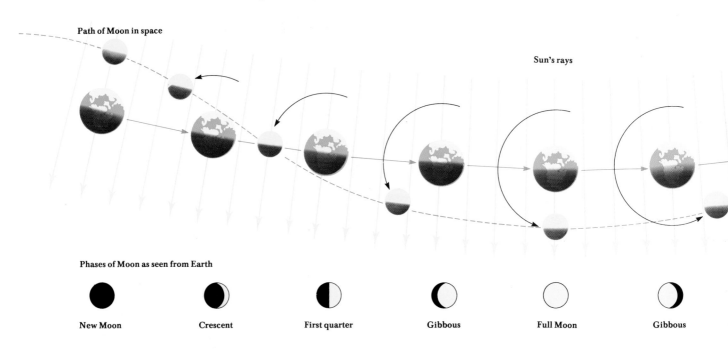

Path of Moon in space

Sun's rays

Phases of Moon as seen from Earth

New Moon          Crescent          First quarter          Gibbous          Full Moon          Gibbous

unquestionable: that the Earth is the centre of all things.

It was not until the 16th century that a Polish monk, Nicholas Copernicus, revived Aristarchus' ideas, and began to argue for a Sun-centred Solar System. Even Copernicus, who brought astronomy out of the Dark Ages, still laboured under some of the misconceptions implanted by the revered classical Greek astronomers and mathematicians: like them, he believed that all motion in the Solar System had to be in perfect circles at uniform speed. Copernicus did not have much observational data to go on, and therefore could theorize freely. He explained the oddities of planetary motions, such as Mars's brief about-turns, by invoking a series of elaborately elegant epicycles, or mini-orbits superimposed on the planets' main orbits.

The true nature of planetary motions was ultimately perceived by Johannes Kepler (1751–1630) who did something that no previous astronomer had done: he tried to fit theories to observations, rather than *vice versa*, and thus created a great watershed in science. He realized that the planets move in elliptical, not circular, orbits, and laid down three great laws of planetary motion which gave rise to Newton's fundamental work on motion and gravitation and thus laid down the foundations of modern physics.

So much for the history. The five 'wanderers' all move in regular elliptical orbits around the Sun. But what about the meteorites and comets? And what about the three other planets, Uranus, Neptune and Pluto, not known to the ancients or to Copernicus or Kepler? These too, all move in elliptical orbits. Everything in the Solar System moves on elliptical orbits. The planets themselves have orbits which are very nearly circular, but the meteorites and comets are much more highly elliptical, sweeping across the Solar System towards the Sun, and then streak-ing away again. Many comets, in fact, have such highly elliptical orbits that they disappear from view on the outer fringes of the Solar System for long periods before returning, and some seem to disappear altogether.

There is one component of the Solar System that we have not yet discussed: the asteroids. None of these is visible to the naked eye, but there are tens of thousands of them. Most are clustered in a well defined belt midway between Jupiter and Saturn, but a few move on highly elliptical orbits which bring them sweeping towards the Sun, across the orbits of Mars and Earth. (There is a remote possibility that one day one of these asteroids could collide with the Earth.)

The complete Solar System, then, consists of nine planets and innumerable meteorites, comets and asteroids gyrating round the Sun in regular elliptical orbits. Each of these members of the Solar System has been studied minutely since the invention of the telescope, but our knowledge of them has advanced far more in the last decade than it did in the previous three millennia, thanks to the advent of spacecraft. Before going on to examine how the different members of the Solar System relate to one another, and how the Solar System as a whole originated, therefore, we shall examine what is known at present of each of the main members.

## The Moon

Although the Moon was mapped through telescopes before its exploration by manned and unmanned spacecraft, we can now interpret its features. Due to photographic missions in lunar orbit, high quality maps are available for practically the whole surface. The samples returned from the American Apollo and Soviet Luna landing sites have completely changed ideas about the origin and early evolution of the solar system, as well as answering many questions about the Moon itself.

The Moon, which has a diameter of 3476 km (2160 miles) orbits the Earth at an average distance of 384,402 km (250,000 miles).

Tracking orbiting spacecraft has shown that the Moon is not perfectly spherical, but is slightly elongated towards the Earth. Earth's tidal forces have locked on to this distortion and caused the Moon to always show the same face to us. This near side is divided into light and dark areas, called by the early investigators *terrae* and *maria* – Latin for lands and seas respectively, since they supposed the surface to be like that of the Earth. Use of these terms has persisted through the years despite more accurate knowledge, although the terrae are now more frequently referred to as *highlands*.

Around the edges of the maria, the highland areas may form mountain chains stretching for hundreds of kilometres. The Appenines reach 7 km ($4\frac{1}{2}$ miles) above the nearby plains. The maria have either circular or irregular shapes and we now know that they are concentrated on the near side.

The almost perfect prints of the sole of an astronaut's boot in the Moon's surface show that the lunar soil is very fine-grained and cohesive, like damp beach sand.

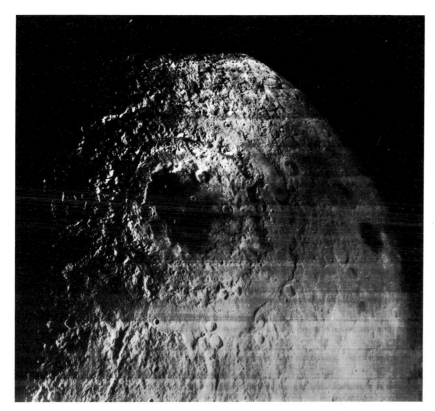

Lunar Orbiter IV photograph of the Mare Orientale basin, and its multiple mountain rings. The outer ring, the Montes Cordilliera, has a diameter of 900 km (560 miles).

## Moon and Earth compared

|  | Moon | Earth |
| --- | --- | --- |
| equatorial diameter (km) | 3 476 | 12 756 |
| sidereal period of |  |  |
| axial rotation | 27·322d | 23h 56m 04s |
| inclination to ecliptic | 1° 32′ | 23° 27′ |
| density (kg per m³) | 3 340 | 5 517 |
| mass (Earth = 1) | 0·0123 | 1·0000 |
| surface gravity |  |  |
| (Earth = 1) | 0·1653 | 1·0000 |
| escape velocity |  |  |
| (km per s) | 2·37 | 11·2 |
| albedo | 0·07 | 0·36 |

mean Earth-Moon distance  384 402 km

Craters have been found on all four inner planets, the Moon and the satellites of Mars, Jupiter, Saturn and Uranus. In the case of the Moon, although present all over the surface, they are most prominent in the highland regions. They are circular features with raised walls and range from large multi-ringed structures with diameters of hundreds of kilometres, all the way down to very small, almost microscopic, pits. The Moon is exceptionally rich in craters with diameters of 20–50 km (12–30 miles) and these, together with the larger sizes, are very shallow in relation to their diameter, with depths of just a few kilometres. Smaller craters tend to be well defined and bowl shaped, but the larger they are, the more likely they are to be partially filled by material which has fallen in from the walls. In the very largest cases there has usually been extensive slumping forming terraces. In some instances, flooding by dark mare material has produced a level floor. Very large craters tend to exhibit a central peak, or even a ring of peaks around the centre.

The origin of craters has been the subject of debate between those who favour a volcanic origin and those who believe they were formed by impact. Evidence from spacecraft data and from the rocks gathered by the Apollo astronauts makes it seem more than likely that the majority are impact features. The impact of a high-velocity meteorite vaporizes both itself and part of the underlying rock down to as much as a few kilometres in depth. The explosion caused by this pocket of hot gas can transport debris to very great distances across the lunar surface. But the crater, produced is always circular, regardless of the direction of the impacting body. Explosion craters on Earth show rims of uplifted bedrock which correspond to the raised walls of lunar craters. Elastic rebound of the rocks can cause the central peaks and peak rings which are seen in the larger, naturally formed craters on both the Earth and Moon. Débris from the explosions can be seen as characteristic *ejecta blankets* and the larger fragments can themselves produce further secondary craters. Earth's gravity makes the ejecta fall close to the main crater, but because the Moon's gravity is only about one-sixth of that of the Earth, relatively young craters on the Moon often show bright rays which sometimes stretch for hundreds of kilometres. These are composed of fine dust and glass beads flung out by the impact. The large multi-ringed structures are thought to have been caused by the impact of very large bodies, perhaps tens of kilometres in diameter.

Ejecta blankets from these impacts can be traced over wide areas of the Moon's surface. It appears that a large number of the earlier structures have been almost obliterated by large numbers of smaller, later craters.

Low-velocity impacts of smaller bodies do not cause explosive cratering, but merely excavate pits by throwing out loose materials. This activity is important in the uppermost layers of the surface, termed the *regolith*, which varies in thickness from 4 to 5 m (13 to 16 feet) on the maria to 10 m (32 feet) and more in the highlands. The uppermost surface layer is composed of the finest dust, but the regolith's composition ranges from this to large blocks several metres across.

In the hostile lunar environment, devoid of air and water, it is impacts which have fractured and powdered the material, and which are responsible for the gradual erosion of craters and other features.

A type of rock which is very common in lunar samples is a *breccia*. This consists of fragments of rock welded together by the heat of later impacts, before being broken up yet again.

There were numerous large impacts in the early part of the Moon's history and some of these excavated basins 20–25 km (12–15 miles) deep. Later on, these basins were flooded by vast quantities of lava which more or less completely filled them, thus forming the circular maria. The irregular maria, on the other hand, have been produced merely by flooding of low-lying terrain and the lava is much thinner.

Domed areas and low arches in the maria have probably been produced by upwelling lavas, but the numerous mare *wrinkle ridges* have almost certainly been caused by compression of the surface layers when the lava flows cooled.

Associated with the edges of the maria are valley-like *rilles*. Sinuous, meandering rilles at first sight look like river valleys. However, unlike water channels they are deepest where they are widest, and in fact they show points of resemblance to collapsed lava tunnels.

Apart from the mare domes which have been mentioned, there are other features thought to be volcanic in origin. Domes on the edges of the highlands seem to have been formed before the maria were filled and may be very ancient, while there are a few low cones which resemble cinder cones produced by low-energy eruptions. More-over, a few dark areas are apparently covered in cinders and ashes from more energetic eruptions. In a number of crater chains the pits greatly resemble the formations known on Earth as volcanic maars, where explosive release of gas has bored a hole in overlying rocks, although there has been no major ejection of volcanic materials. All

these features are of minor importance, however, and the few large craters which on the grounds of their positions and associated features may be volcanic, are greatly outnumbered by the impact formations. Indeed, although the Apollo 16 landing site of Descartes was chosen because the surface rocks were possibly volcanic, in the event the astronauts collected large quantities of impact breccias.

The occasional obscurations and glows which are known as *transient lunar phenomena* are most frequently seen around the edges of maria and near to certain relatively fresh craters. These events are more numerous when the Moon is at perigee, suggesting that tidal forces are causing slight movement of the crust which permits gas to escape from the interior. Gas emissions have been detected from Earth, and more reliable observations have been made from orbiting spacecraft, but there are no reasons for supposing that any eruptive volcanic activity involving release of new material to the surface is taking place.

Seismometers placed by the Apollo astronauts have recorded waves from moonquakes as well as from the impact of natural bodies and spent spacecraft stages. Over 3000 natural moonquakes have been recorded per year. The majority of these

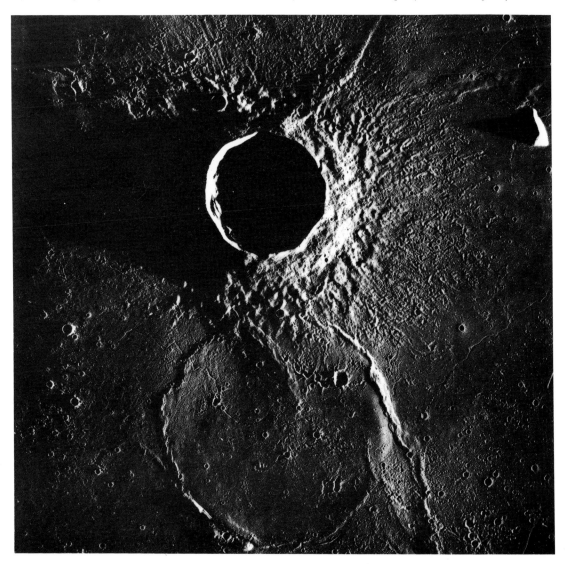

The crater Lambert in the Mare Imbrium, showing the radial structure of the ejecta blanket and also many secondary craters. The mare ring structure to the south has a diameter of about 50 km (30 miles) and seems to have been caused by lava flows covering, and then subsiding onto an earlier crater wall.

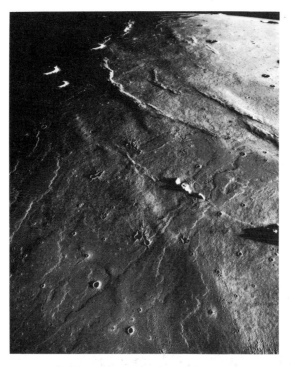

Lava flows in the Mare Imbrium which are here about 35 m (120 feet) high and 10–25 km (6–15 miles) wide. The lava source was off the picture to the lower left and some flows are about 1200 km (750 miles) long.

occurred at very great depths (about 900 km or 560 miles), with very few in the topmost crustal layer. More moonquakes occur when the Moon is at perigee, once more suggesting that tidal forces are important.

In contrast to the Earth, whose molten core sometimes bubbles to the surface in the form of volcanoes, the Moon is rigid down to a depth of about 1000 km (600 miles), below which there is a weak, possibly molten, zone. The temperature at the centre is probably in the region of 1300 K. As the overall density of the Moon is much less than that of the Earth (3340 kg per m³ against 5517 kg per m³) it cannot have a very large metallic core of iron and nickel, and the centre is probably composed of a mixture of iron and sulphur, with perhaps a very little nickel. Returned rock samples indicate that the Moon had a magnetic field a very long time ago, although there is none now, and this could have been produced by such a composition when most of the interior was fluid.

Analysis of the motion of spacecraft in orbit around the Moon, particularly the Lunar Orbiter series of craft, showed certain gravity anomalies in some regions. These mass concentrations, or *mascons*, were found under the circular maria in particular. Their formation is still a subject of debate but it would appear that some basins formed by impacts became the sites for a concentration of denser material either at the surface or at depth.

There are various dating techniques which are used on lunar samples and these have shown that the majority of the rocks are very old. The highland materials, for example, commonly have

ages of $4.0$–$4.3 \times 10^9$ years and the very oldest rock has been dated at $4.6 \times 10^9$ years, very close to the age estimated for the solar system itself. The mare basalts on the other hand have ages of $3.1$–$3.8 \times 10^9$ years and are thus much younger than the highlands, although still comparable in age to the very oldest rocks known on Earth ($3.8 \times 10^9$ years). No significantly younger rocks are known to exist on the Moon – it is very old indeed.

It is thought that the Moon (like all the planets) was formed in a very short time from the accretion of smaller bodies a few hundred kilometres in diameter, which are known as planetesimals. In the intense early bombardment of the planetesimals, the heat of the impacts caused the whole surface to become molten down to about 200 km (120 miles). From this molten layer the highland rocks formed, and these continued to be cratered after they solidified.

At a later date, the interior heated up due to radioactivity of the materials forming its interior and the mare lavas escaped to the surface, filling some of the large basin which remained. The interior of the Moon gradually cooled until now only the very centre retains any heat. Being without a molten interior today, the Moon does not possess any significant magnetic field. There are, however, traces within its rocks of very early magnetism, which probably did originate within the Moon itself.

*Above*: The moon, dotted with craters formed by meteorites, shows the distinctive dark *maria* (seas) – in fact, the most massive craters of all. The Moon keeps one face turned towards the Earth. Soon after its formation, the Earth's gravity created tidal forces in its rocks which slowed its rotation by friction until it revolved on its axis once for every orbit around the Earth. *Maria* are largely absent from the far side and are thought to be related to the Moon's orientation.

A composite of two photographs taken on the lunar surface at the Apollo 17 landing site in the Taurus Mountains near the crater Littrow. The valley is thought to be a graben flooded with basalt flows totalling about 1400 m (4500 feet). Astronaut Schmidt and the Lunar Rover give an idea of the scale of the surface features, while South Massif on the right is about 8 km (13 miles) distant and reaches a height of 2500 m (8000 feet). The large broken boulder is a breccia, and has rolled about 1·5 km (2·5 miles) down the slope.

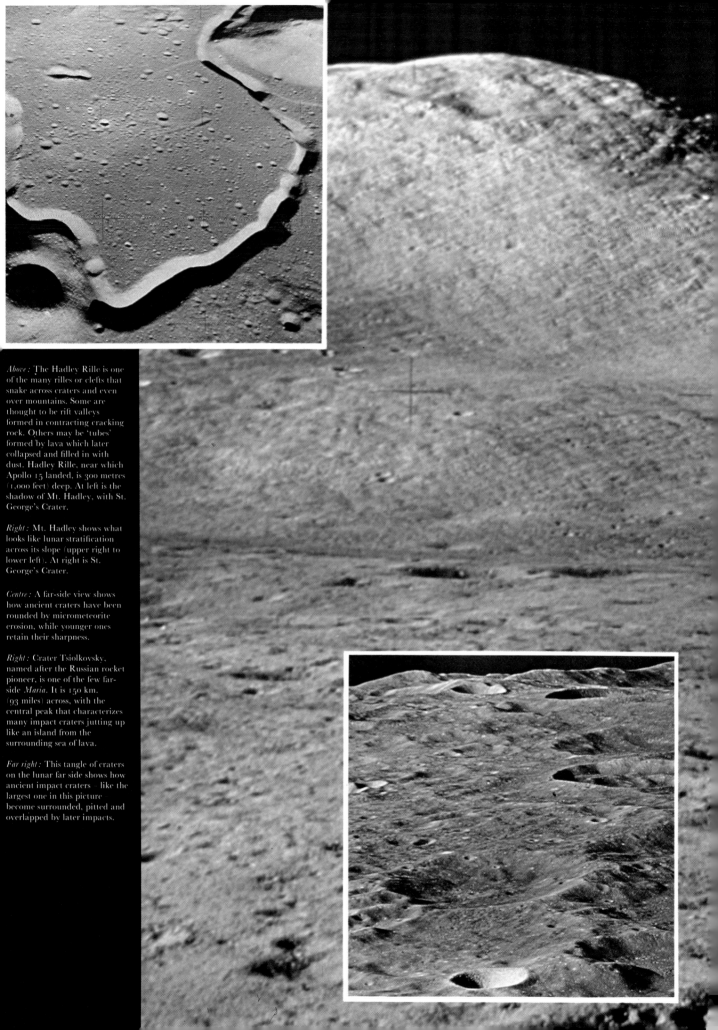

*Above:* The Hadley Rille is one of the many rilles or clefts that snake across craters and even over mountains. Some are thought to be rift valleys formed in contracting cracking rock. Others may be 'tubes' formed by lava which later collapsed and filled in with dust. Hadley Rille, near which Apollo 15 landed, is 300 metres (1,000 feet) deep. At left is the shadow of Mt. Hadley, with St. George's Crater.

*Right:* Mt. Hadley shows what looks like lunar stratification across its slope (upper right to lower left). At right is St. George's Crater.

*Centre:* A far-side view shows how ancient craters have been rounded by micrometeorite erosion, while younger ones retain their sharpness.

*Right:* Crater Tsiolkovsky, named after the Russian rocket pioneer, is one of the few far-side *Maria*. It is 150 km. (93 miles) across, with the central peak that characterizes many impact craters jutting up like an island from the surrounding sea of lava.

*Far right:* This tangle of craters on the lunar far side shows how ancient impact craters – like the largest one in this picture – become surrounded, pitted and overlapped by later impacts.

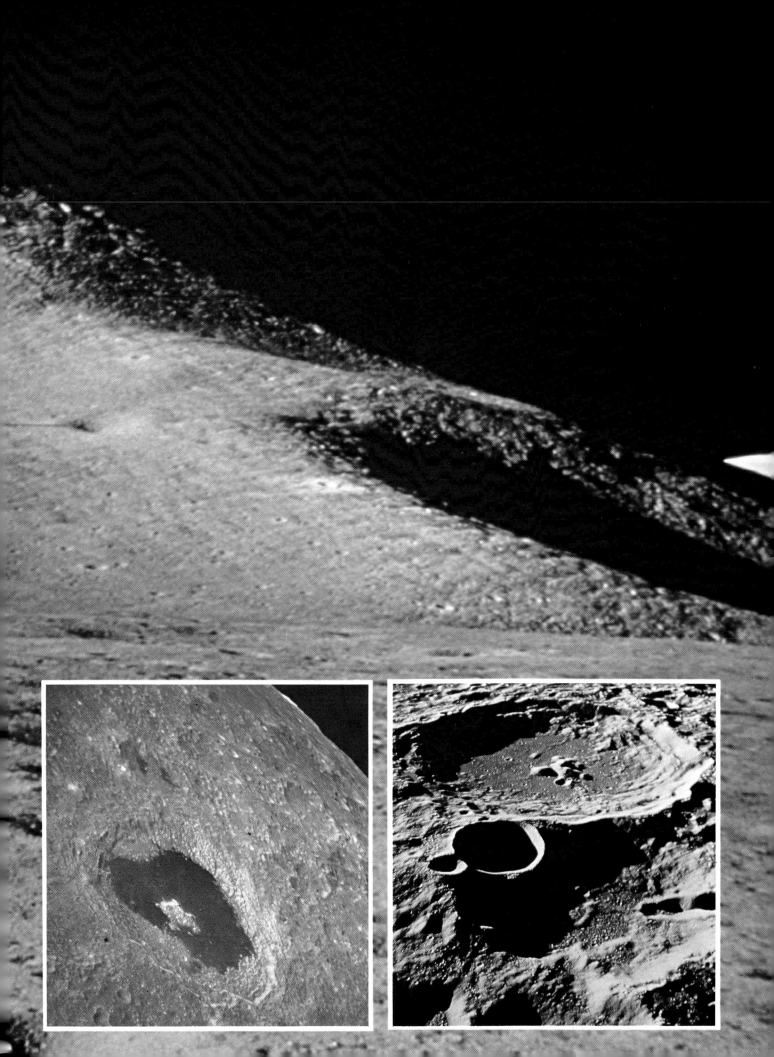

## Mercury

Mercury is not an easy object to observe from the Earth because of its close proximity to the Sun. It never reaches an elongation greater than about 27°, and it is also very small with a diameter of 4878 km (3000 miles). Only a few vague markings are visible with large telescopes. It was long supposed that its axial rotation was the same as its orbital period of 88 days, so that one hemisphere permanently faced the Sun, leading to very high temperatures on that side while the other was always cold. Using radar the rotation period was found to be approximately 59 days, suggesting that tidal interaction with the Sun has caused 2 orbital periods to equal exactly 3 axial rotations. This effect, known as *spin-orbit coupling*, has resulted in a rotation period of 58·65 days. The Mariner 10 spacecraft, to which we owe practically all knowledge about Mercury, was placed into a similar resonant orbit, making its second and third encounters with the planet 2 and 4 Mercurian "years" (3 and 6 rotations) after its initial approach.

Mercury has a very high density, greater than any of the other planets except the Earth. This is surprising in such a small body and suggests the presence of a relatively large metallic iron-nickel core, which contains about 80 per cent of the planet's mass (as compared with the Earth's 32 per cent). This was confirmed by the Mariner observations which showed Mercury to have a magnetic field, a fact which presumably indicates that the planet has a fluid core. The field strength is weaker than that of the Earth. The exact position of the magnetic axis is unknown, but it is thought to coincide with the rotational axis, probably at right-angles to the orbital plane.

The very high density of the planet remains a mystery, but although some doubts have been expressed about past measurements, the results obtained from the Mariner 10 tracking showed that the density is correct.

Mercury has a very tenuous atmosphere which consists mainly of sodium. In addition there are helium and hydrogen atoms, captured from the solar wind, and retained for about 200 days before

### Mercury and Earth compared

|  | Mercury | Earth |
|---|---|---|
| equatorial diameter (km) | 4 878 | 12 756 |
| sidereal period of | | |
| axial rotation | 58·65d | 23h 56m 04s |
| inclination to orbit | 0°? | 23° 27′ |
| density (kg per m³) | 5 500 | 5 517 |
| mass (Earth = 1) | 0·055 | 1·0000 |
| surface gravity | | |
| (Earth = 1) | 0·38 | 1·0000 |
| escape velocity | | |
| (km per s) | 4·3 | 11·2 |
| albedo | 0·06 | 0·36 |

mean Sun-Mercury distance 0·3870987 AU

they gain sufficient energy to escape back to space.

Although the surface temperatures are not so extreme as had been thought prior to Mariner 10, they can reach 700 K at the equator of the sunward hemisphere and cool to less than 100 K (−173°C) on the dark side. Due to the orbital coupling either longitude 0° or 180° is towards the Sun when the planet is at its closest, at perihelion, while longitudes 90° or 270° face it at aphelion. As a result of the various motions and the orbit's great eccentricity, the 0° and 180° meridians receive about two-and-a-half times as much solar radiation as those at 90° and 270°.

Mariner 10 transmitted back a number of pictures which showed the surface of the planet to be covered with a large number of craters. It was only possible to examine a little more than one-third of the surface, but, as on the Moon, there is a division into highlands and lower mare-like areas. The highlands are not as saturated with 20–50 km (12–30 miles) craters as the Moon and there are extensive relatively flat regions known as *intercrater plains*. It has been suggested that these are the original Mercurian surface which has undergone a lesser degree of cratering than the Moon, but close examination shows evidence of a large number of highly degraded craters in the plains. It therefore seems possible that the surface has gone through a process of heating and softening, similar to that which formed the Moon's original crust. Since the crust became completely rigid, insufficient impacts have occurred to cover the surface with craters.

The craters resemble those on the Moon, but secondary craters are closer to the main feature and ray systems are less extensive. This is due to the higher surface gravity. Central peaks and small diameter peak rings are also present.

In summary, it seems that the history of Mercury has been very similar to that of the Moon, with crustal heating and one or more major episodes of impact cratering. Crater counts suggest that Mercury, the Moon and Mars have been affected by a similar meteoroid flux in the relatively recent past.

## Venus

Despite the fact that Venus approaches closer to

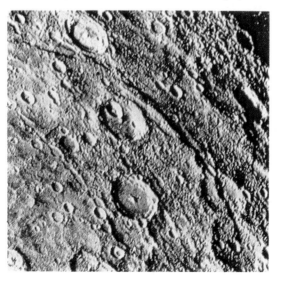

Mariner 10 photograph of part of the surface of Mercury. The prominent lobate scarp which runs diagonally across the picture is actually considerably longer, only about 300 km (190 miles) being shown here.

the Earth than any other major planet and that it is very similar in size and total mass to the Earth and the several Mariner, Pioneer Venus and Venera spacraft missions, we know less about its surface features than those of any other body of the inner solar system.

Venus has a very extensive atmosphere with an albedo (surface reflectivity) of 76 per cent, and this completely hides the surface. Even the rotation period could not be established with any confidence until 1962, when radar methods indicated a retrograde period of 243 days – Venus rotates on its axis in the opposite sense to the Earth, and its "day" is longer than its year! The indistinct markings, sometimes visible from the Earth and on Mariner 10 photographs, show an apparent 4-day rotation period for the upper atmosphere; this is discussed below.

The planet's overall density is fairly close to that of the Earth and it would be reasonable to assume that both planets had a similar composition when they were formed. The implication of this is that the core of Venus has a radius of about 3100 km (1900 miles) with a considerable proportion being fluid. It is expected that there is a mantle and a crust which probably closely resemble those of the Earth. In spite of there being a suspected fluid core, the planet has no detectable magnetic field, the rotation apparently being too slow to produce one.

Problems have arisen since the discovery of anomalies in the abundances of certain gases.

Radar can penetrate the dense atmosphere. Studies using the large Arecibo telescope and the orbiter of Pioneer Venus have revealed considerable surface detail. (The orbiter radar had low resolution of 100 km (60 miles) whereas Earth-based radar can approach 6 km (4 miles) in certain regions.) The main portion of the planet is surprisingly flat. High regions include Aphrodite Terra, Beta Regio (Thea Mons and Rhea Mons) and Ishtar Terra. Maxwell Montes rises to 11 km (7 miles) above the mean radius of 6051·2 km (3760 miles).

### Venus and Earth compared

|  | Venus | Earth |
|---|---|---|
| equatorial diameter (km) | 12 104 | 12 756 |
| sidereal period of axial rotation | 243·16d | 23h 56m 04s |
| inclination to orbit | 178° | 23° 27′ |
| density (kg per m³) | 5 250 | 5 517 |
| mass (Earth = 1) | 0·815 | 1·0000 |
| surface gravity (Earth = 1) | 0·903 | 1·0000 |
| escape velocity (km per s) | 10·36 | 11·2 |
| albedo | 0·76 | 0·36 |
| mean Sun-Venus distance | 0·7233322 AU | |

This photomosaic of Mercury was constructed of eighteen photos taken at 42-second intervals by Mariner 10 six hours after the spacecraft flew past the planet on March 29 1974. The north pole is at the top and the equator extends from left to right about two-thirds down from the top. A large circular basin, about 1300 km (800 miles) in diameter, is emerging from the day-night terminator at left centre. Bright-rayed craters are prominent in this view of Mercury. One such ray seems to join in both east-west and north-south directions. Taken from a distance of about 210,000 km (130,000 miles), the pictures were computer-enhanced at the Jet Propulsion Laboratory.

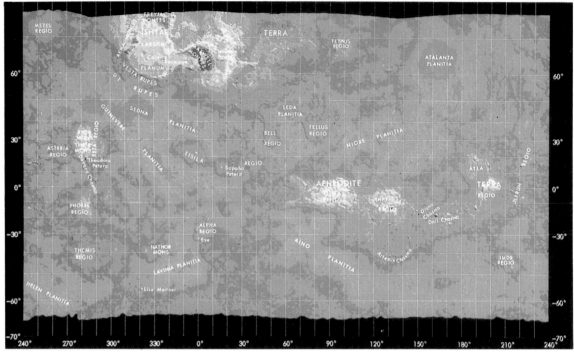

A radar map of the surface of Venus, made by the Orbiter section of the Pioneer Venus spacecraft.

Many of the landforms are similar to terrestrial shield volcanoes, and high-resolution images of Maxwell Montes appear to show a 100 km (60 mile) summit crater and other volcanic features. The equatorial features, especially Beta Regio, show radio bursts. These appear to be ionospheric noise rather than electrical discharges in volcanic ejecta.

Aphrodite Terra appears to contain aligned troughs similar to features at the Earth's mid-ocean ridges; this could be evidence of plate-tectonic activity. There are indications of large basins and a considerable number of craters, ranging from about 160 km (100 miles) to 35 km (22 miles) in diameter, and perhaps larger impact structures, including one 1800 km (1100 miles) in diameter. There are few small craters, probably because the dense atmosphere prevents small meteoroids from reaching the surface. Surface winds are now known to be very light, so that any erosion of surface rocks is likely to be by chemical and thermal processes rather than by wind-blown particles.

Measurements of the radioactivity of surface rocks were made by the Soviet Veneras 8, 9 and 10. At the first site the rock was like terrestrial granites, and at the others resembled both terrestrial and lunar basalts. The more sophisticated techniques used on Veneras 13 and 14 and Vega 2 (using drilling and X-ray fluorescence) found an unusual high-potassium basalt, with some similarities to the lunar highlands and terrestrial continental rocks, a basalt similar to that of the Earth's ocean floors, and an anorthositic rock resembling lunar highland material and some very ancient terrestrial rocks. Veneras 9, 10, 13 and 14 returned pictures of their landing places (the last two in colour). The first, in particular, proved to be surprisingly rough; the others were generally smoother, suggesting that erosional processes are at work. Veneras 10 and 13 showed flat rock outcrops with intervening patches of darker "soil" fragments. The view from Venera 14 was somewhat

different, with just a flat, apparently layered rock expanse, but the layers are rather difficult to understand and have been proposed as either cemented fine particles or thin layers of the original lava.

The main constituent of the atmosphere is carbon dioxide, which amounts to 97 per cent of the total. The overall quantity of carbon dioxide is approximately the same as the total held by the combined oceanic, rock and atmospheric reservoirs on Earth, implying that both planets were formed with, or have accreted, similar quantities of the gas. A considerable number of other components of the atmosphere have now been identified. Particularly significant is the very small amount of water and this will be discussed below.

Because of the vast amount of carbon dioxide, the atmospheric pressure is about ninety times that of the Earth. As a result of the carbon dioxide's "greenhouse effect", the surface temperature has been raised to 760 K, comparable to, if not hotter than, that at the surface of Mercury. However, there is now considerable evidence that these high temperatures have not prevailed throughout the planet's history.

The extreme surface temperature is, of course, far above the boiling point of water, and it was a matter of some surprise when early radio studies showed that the atmosphere contains very little water vapour. Subsequent measurements agree with this. This low measure is difficult to understand unless Venus was formed from very different material than the Earth, or gained its atmosphere in a different way. But there is direct evidence that Venus once had a considerable quantity of water. This information comes from a determination of the ratio of hydrogen to deuterium (heavy hydrogen). There is far more deuterium on Venus than on Earth, and this shows that a large amount of water was once present, indeed that oceans probably existed. The greenhouse effect would cause this liquid water to be evaporated into the atmosphere (incidentally contributing to a still stronger trapping of infrared radiation), and the water molecules would be dissociated by the Sun's ultraviolet radiation at the top of the atmosphere. The ordinary light hydrogen atoms would be most easily lost to space, but the heavier deuterium isotope would be retained, leading to its present high concentration. The free oxygen is likely to have been incorporated into surface rocks, and it has been suggested that the oxidation process would further increase the heating of the crustal layers. However, it does seem likely that Venus had large bodies of water at an early period in its history, and it is by no means impossible that some form of life did originate upon the planet before the greenhouse effect made it too hot for any life forms.

The composition of the clouds was a complete mystery, but now it's quite well understood. The yellowish tinge is probably due to the presence of sulphur, while the dark markings, which only show up in ultraviolet light, are most probably caused by absorption by sulphur dioxide. The clouds exhibit a strongly layered structure, the densest layers occurring between about 70 and 47·5 km (44 and 30 miles), with hazes both above and below. The lower atmosphere is essentially clear, which is one of the reasons why conventional lightning processes are unlikely to occur.

There are various sizes of cloud particle present, but the majority appear to be droplets of sulphuric acid, some tiny and some larger, the latter accumulating on small particle nuclei. The very large crystals, most likely to be some form of chloride, although this is still uncertain. A number of other components must be present to account for the optical properties of the clouds.

The atmospheric circulation is unusual. There is a high-speed circulation of the upper and middle layers around the planet, giving rise to an apparent rotation period of about 4 days for the cloud tops.

The Soviet Venera and Vega landers and the four US Pioneer Venus probes (one of which, although not designed to survive impact with the surface, did return data for 67 minutes) show that information can be gained slowly about the nature of the surface, as well as about the complicated structure of the atmosphere. Better knowledge of the planet's various features should come with the Magellan radar-mapping mission, scheduled for launch in 1989, and perhaps future, long-lived versions of the simple balloon probes deployed by the Vega 1 and 2 missions, each of which lasted for 42 hours.

## Mars

Mars, unlike Venus, has only a thin atmosphere and has long been studied from Earth. Detailed maps of its surface markings have been drawn. These were thought to be related to high and low areas, perhaps somewhat similar to those found on the Moon. Apart from these features it shows clouds, brilliant polar caps which change with the seasons, similar seasonal changes in some of the surface markings and occasional vast dust storms which are frequently planet-wide. Rather ironically, the markings which were thought to indicate the nature of the surface have now been shown by spacecraft pictures to bear little relation to actual features, whereas the other characteristics have been fully confirmed.

Mars has a lower density than any of the other inner planets and has a surface gravity almost exactly equal to that of Mercury, despite being more than 3800 km (2360 miles) greater in diameter. The crustal and mantle materials of the inner planets appear to be very similar and this, together with spacecraft tracking results, suggests that the core of Mars is smaller and less dense than those of the other planets, with the most probable material being iron sulphide. The nature of the core may be clarified by future information about the planet's magnetic field. Results from the Soviet Mars probes and the US Viking spacecraft appear to indicate a weak magnetic field, although nothing had been detected by the Mariner 4, 7 and 9 missions.

Both the Mariner 4 and 7 spacecraft showed that the surface of Mars was cratered, but it was Mariner 9 which revealed the true distribution of the craters and discovered numerous other interesting features. The Soviet Mars 5 probe also returned high resolution pictures, but the highest quality images were obtained by the two Viking orbiters. The most striking fact to be revealed is that the surface is effectively divided into two hemispheres, one of which, the southern, is generally high, heavily cratered and ancient, while

A photograph of the surface of Venus taken by Venera 10. This has been processed to correct the original distorted shape produced by the camera. The horizon line is marked. The flat surfaces in this photograph suggest that erosional forces have been at work.

the northern is formed of low-lying, relatively featureless plains which have far less, and generaly fresher craters. These plains appear to be predominantly formed of volcanic materials and this general impression was confirmed by the Viking 2 lander. The boundary between the two regions is marked by boundary scarps which are remarkably uniform in height, and areas where the highlands are being eroded.

Craters in the northern hemisphere only rarely exceed 10 km (6 miles) in diameter, but in the southern they include several multi-ringed structures which range in size up to that of Hellas, which is about 2000 km (1200 miles) across and some 4 km (2½ miles) deep. The number of central peaks

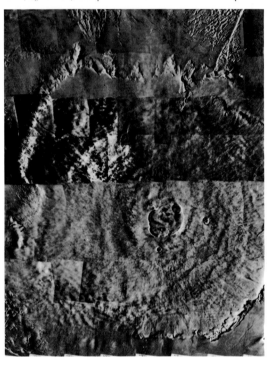

and peak rings is similar to that of Mercury, as expected from the almost equal surface gravities. The effect of gravity can also be seen in the fact that Martian craters are shallower than those on the Moon. Distinct ejecta blankets can be seen around some of the younger impact craters.

Although there are no signs of curved mountain chains, a prominent part of the surface is that in the three volcanic regions in the Tharsis, Elysium and Hellas areas. One of these, the Tharsis volcanic province, is especially notable as it contains extensive volcanic plains, four exceptionally large volcanoes and numerous other features. The largest volcano, Olympus Mons, has a diameter of 600 km (370 miles) and a height of 26 km (16 miles), making it twice as high as the large feature on Venus and much bigger than Hawaii, which forms the largest such volcano on Earth. It has a complex central caldera, has evidence of extensive lava flows, and a boundary cliff which is up to 4 km (2½ miles) high in places.

The other three volcanoes in the same area are very large with diameters of about 400 km (250 miles) and heights of 19 km (12 miles) but there are

also smaller domes which were probably formed by more viscous lava. To the north is a degraded older structure called Alba Patera, which seems to have had a diameter of about 1600 km (1000 miles), which makes it the largest volcanic feature on the planet.

The Elysium region contains volcanoes and volcanic plains, but the structures in the Hellas region are much less distinct and certainly very much older.

The Tharsis Plateau volcanoes lie on part of a very large uplifted area, which is about 5000 km (3000 miles) across and 7 km (4 miles) high. The exact causes of this bulge are obscure. The immense Vallis Marineris canyon system is several kilometres deep and extends for about 4000 km (2500 miles) across the planet, roughly parallel to the equator. At its eastern end it joins the Chryse Trough which runs south to north across the cratered upland terrain.

Present conditions do not permit rainfall or areas of liquid water on Mars. Some features may have been cut by water flowing beneath an insulating blanket of snow and ice, or by violent, short-lived floods that burst out from beneath the ice. But some of the channels appear to have formed on the surface, and sediments in Vallis Marineris have been interpreted as ancient lake deposits.

The sediments are difficult to explain unless open bodies of water existed. The early atmosphere was probably much denser and warmer, and consisted mainly of carbon dioxide, which was subsequently trapped in carbonate rocks. With the loss of atmospheric carbon dioxide, the temperature declined and the water became locked into the regolith or thick layers of permafrost, with a small quantity in the polar caps. Much of the water was broken down into hydrogen (which escaped to space) and oxygen (which became incorporated into the surface rocks).

Due to the low atmospheric density, higher wind speeds are needed on Mars to transport particles of a similar weight than on Earth. The Viking Landers actually recorded speeds of only 3–6 m per second (6–12 miles per hour) on the surface, but found much higher velocity higher in the atmosphere.

Photomosaic of Olympus Mons, which has a diameter of about 600 km (370 miles) and height of 26 km (16 miles). The complex central caldera has a diameter of about 70 km (44 miles).

### Mars and Earth compared

|  | Mars | Earth |
|---|---|---|
| equatorial diameter (km) | 6 794 | 12 756 |
| sidereal period of axial rotation | 24h 37m 23s | 23h 56m 04s |
| inclination to orbit | 24° 46′ | 23° 27′ |
| density (kg per m³) | 3 933 | 5 517 |
| mass (Earth = 1) | 0·107 | 1·0000 |
| surface gravity (Earth = 1) | 0·38 | 1·0000 |
| escape velocity (km per s) | 5·03 | 11·2 |
| albedo | 0·16 | 0.36 |
| mean Sun-Mars distance | 1·5236915 AU | |

The planet-wide dust storms seem to originate in the southern hemisphere, but there is a strong tendency for the fine particles to be removed from the high cratered terrain and to be deposited on the lowlands, particularly on the northern plains. The action of wind-borne particles can be seen in large areas of grooved and fluted surfaces all over the planet.

The Viking Landers measured the average surface pressure as $7 \cdot 4 \times 10^{-3}$ atmospheres and that the mean temperature as 230 K, which is 43°C below the freezing point of water.

The main constituent of the Martian atmosphere is carbon dioxide. It is difficult to establish the amount of carbon dioxide and water which have been produced by loss of trapped gas on Mars, but the quantities seem to have been many, possibly hundreds of times less than on Earth. The permanent polar ice caps and the majority of the clouds and fogs are formed of water ice. There has long been a debate concerning whether temperatures are low enough for carbon dioxide to freeze, but spacecraft measurements indicate temperatures as low as 125 K, so that the seasonal caps are probably formed of both water and carbon dioxide ices. Carbon dioxide ice is also present in some

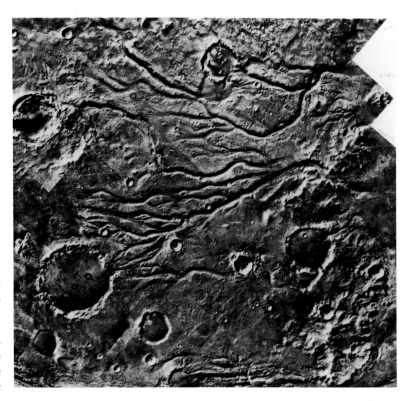

*Above*: Photomosaic of the channelled terrain west of the Viking Lander 1 site in Chryse Planitia. The slope is from west to east (left to right). Note that the channels cut some craters, but also have other, later craters superimposed upon them.

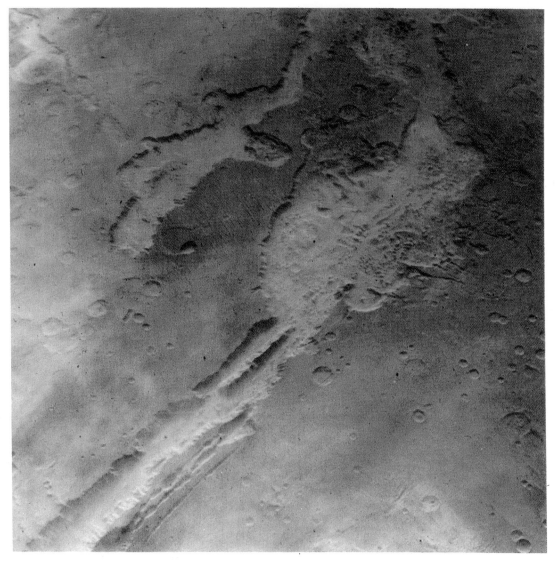

Viking Orbiter 1 photomosaic of Vallis Marineris, the area covered being about 1800 km (1100 miles) by 2000 km (1240 miles). The canyon system is 4000 km (2480 miles) long and up to 120 km (75 miles) wide and 6 km (4 miles) deep. Note how the depressions are extending along lines of structural weakness (bottom centre). General levels trend downwards towards the Chryse Trough which is out of the picture to the upper right.

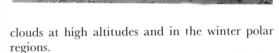

clouds at high altitudes and in the winter polar regions.

Mars possesses two natural satellites, Phobos and Deimos. Phobos, the larger, orbits the planet in 7·65 hours at a distance of 2·75 Mars radii. Deimos, the smaller, orbits in 30·3 hours at 6·9 Mars radii.

Tidal forces acting on these irregular bodies have pulled both satellites into synchronous rotation so that the same faces are always turned towards Mars.

Both are covered with craters, the largest on each being Stickney (diameter 10 km (6 miles)) on Phobos, and Voltaire (diameter 2 km (1¼ miles)) on Deimos. The craters show uplifted rims but no ejecta blankets or central peaks due to the low

*Above left*: A composite of three single-colour photographs of Mars made by Viking Orbiter 1. Part of the south polar cap can be seen, while frost covers the area between it and Argyre basin, which is itself beneath frost or haze. Vallis Marineris is recognizable towards the top, but water ice clouds cover the Tharsis volcanic region.

*Above right*: In reconstruction, Olympus Mon's cone shape – leading up to the 65 km (40 mile) wide caldera – forms a broad shield, somewhat similar to those of the Hawaiian volcanoes.

surface gravity. The sharpness of some of the features suggests that both bodies are solid, rather than loose blocks bound together by gravitational forces.

The Soviet Union launched two spacecraft to Mars in July 1988, in order to learn more about Phobos. The craft, Phobos 1 and 2, were identical and were to descend to within 50 m of Phobos itself. Each was to have fired laser and ion beams at the tiny, mis-shapen moon, in order to vapourize small amounts of the surface for analysis by instruments aboard the craft. Four probes would have been dropped to the surface to return data for over a year. However, contact with Phobos 1 was lost in August 1988. Phobos 2 entered Martian orbit in January 1989 and returned high-quality data until March, when all contact with the craft was abruptly lost. Clearly, this was a major blow to planetary astronomers who must now wait for a future mission to reveal the secrets of the Martian moon.

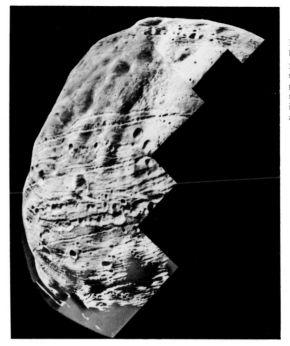

Photomosaic of Phobos, taken by Viking Orbiter 1 when about 300 km (180 miles) from the satellite. The striking linear grooves are about 500 m (0·3 miles) wide. The crater Stickney is indistinctly seen on the limb at top left.

This image, returned by Viking 1 orbiter, reveals the thin carbon dioxide atmosphere that envelopes Mars. Several high-altitude cloud layers are visible on the horizon. The large basin in the foreground, named Argyre Planitia, was left by an asteroid impact.

*Opposite*: The boulder-strewn field of red rocks reaches to the horizon, 3 km (1.9 miles) from Viking 2 on Mars' Utopian Plain. Scientists believe the colours of the Martian surface and sky in this photograph represent their true colours. Note the blue starfield and red stripes of the flag. Fine particles of red dust have settled on spacecraft surfaces. The salmon colour of the sky is caused by dust particles suspended in the atmosphere. The circular structure at top is the high-gain antenna, pointed toward Earth. Viking 2 landed on September 3, 1976, some 7400 km (4600 miles) from its twin, Viking 1, which touched down on July 20.

In March 1979 Voyager I revealed startling details of Jupiter, a gaseous giant twice the mass of all the other planets combined, and of its four major inner satellites. *Top*: From 47 million km (29 million miles) Jupiter's banded atmosphere and two satellites, Ganymede and Europa, are clearly visible. *Above left*: Europa is a bright, icy moon a little smaller than our own and Callisto (*above right*) is dark and strangely mottled.

## Jupiter

Ever since Galileo Galilei turned his primitive telescope towards Jupiter in 1609 and noted its four major satellites, the planet has been under constant study. It has proved to be a world of superlatives: it is by far the largest of the planets with an equatorial diameter of 142,796 km (89,000 miles) (11·2 times that of the Earth); it is more massive than all of the other planets put together (318 times the Earth's mass); it has the shortest rotation period; it has a vast magnetic field and is a powerful source of radio waves. It has at least seventeen satellites and exerts great influence on the orbits of the minor planets and comets by its gravitational perturbations. However, it is small when compared with the Sun which has a mass 1047 times that of Jupiter.

Despite its great mass, its density is only a quarter that of the Earth, and this is a clear guide to the nature of its interior. Beneath an atmospheric layer about 1000 km (625 miles) thick there is probably a layer of liquid hydrogen, and below that a core of liquid *metallic* hydrogen. At the very centre of the core there may be a small kernel of rocky, silicate material, but even this is not certain. It does seem likely, however, that there is nothing that one could describe as a solid "surface" to the planet.

The only parts of the planet that are accessible to observation are the extreme outer layers of the atmosphere, which consist mostly of ammonia and hydrogen with smaller amounts of methane and water. Even a modest telescope will reveal a series of dark bands or stripes running across the yellowish disc of the planet; these are interpreted in terms of circulating currents of gases sweeping along parallel with the equator, but in different directions. It is certain that enormously powerful winds blow in the upper atmosphere of Jupiter.

In part, this is due to the extremely rapid rotation of the planet, which revolves on its axis in only nine hours and 50 minutes. This rapid rotation is also responsible for another characteristic feature of Jupiter: it is visibly flattened, so that it looks a little like a squashed orange. The centrifugal force induced by the rapid rotation forces material outwards in the equatorial regions, causing this to bulge relative to the polar regions.

Another of Jupiter's most famous features is also an indirect result of the rapid rotation. The powerful counter-currents of gases set up in the upper atmosphere give rise to long lasting spiralling systems, a little like the cyclones which are well known in tropical regions of the Earth. Jupiter's Great Red Spot, originally interpreted as a great volcano rearing above the surface, is almost certainly one of these circulating atmospheric disturbances, although the reason for its red colour is not fully understood. The Great Red Spot has been continuously visible since it was first discovered by Giovanni Domenico Cassini in 1665; no other single feature has been visible for as long.

The magnetic axis is tilted by about 10–11° from the axis of rotation and displaced by about 7000 km (4300 miles) from the centre of the planet.

Although the Pioneers succeeded in obtaining some superb pictures of the details of Jupiter's atmosphere, and the Voyagers obtained even more – even recording the existence of a thin, diaphonous ring system – much of the interest in the two missions lies in the data on Jupiter's numerous moons.

There are no less than 16 satellites, and more may well be discovered by future spacecraft. The

### Jupiter and Earth compared

| | Jupiter | Earth |
|---|---|---|
| equatorial diameter (km) | 142 796 | 12 756 |
| sidereal period of axial rotation | 9h 55m 30s | 23h 56m 04s |
| inclination to orbit | 3° 04′ | 23° 27′ |
| density (kg per m³) | 1 330 | 5 517 |
| mass (Earth = 1) | 1318·7 | 1·0000 |
| surface gravity (Earth = 1) | 2·643 | 1·0000 |
| escape velocity (km per s) | 60·22 | 11·2 |
| albedo | 0·73 | 0·36 |
| mean Sun-Jupiter distance | 5·2028039 AU | |

interest in them lies in the fact that many of them are large bodies in their own right, and some are more solid and rockier than Jupiter itself. The four largest satellites, Io, Europa, Callisto and Ganymede, are known as the Galilean satellites, since they were discovered by the great Galileo himself, as soon as he turned his first telescope towards the planet. Of these four, Ganymede and Callisto are the biggest, with diameters of over 5000 km (3125 miles). They are therefore bigger than the planet Mercury. There is a considerable range in density amongst the four, ranging from Io (about 0·65 Earth's), to Callisto (about 0·3 Earth's). Thus it is thought that Io may be made up predominantly of rocky material closely similar to our own Moon, while Callisto and the other low-density satellites are probably composed of ice and various rocky materials.

The Galilean satellites, then, are large and interesting bodies in their own right. They have an additional attraction, however, in that since it is impossible to land directly on Jupiter, one of the satellites would make an excellent place to set up an observatory to study Jupiter. It will probably be many decades before this can be accomplished, but the technical difficulties are not very great. It will be a fascinating stage in the exploration of the Solar System.

The US Galileo mission to Jupiter, scheduled for launch in October 1989, is expected to reveal a great deal of new information about major moons as well as the structure of the Jovian atmosphere.

Voyager 2 photograph of Jupiter taken on June 28 1979. One of the dark 'barges' can be seen in the North Equatorial Belt. In the lower half of the picture, south of the Equatorial Zone, a chaotic region of whiter clouds is visible, lying west of the Great Red Spot, which is out of the picture to the right.

Unlike Jupiter, Saturn has a thick overlying haze which hides most of the atmospheric features. Complete computer-processing and image-enhancement was required before any details could be seen, as in the case of this Voyager 2 picture.

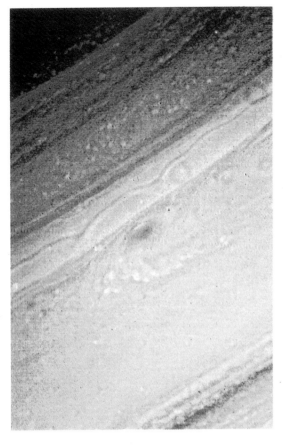

## Saturn

The most spectacular object in the solar system as seen through a telescope is probably Saturn with its magnificent system of rings. Apart from these, however, Saturn shows many similarities with Jupiter, and is a very sizeable planet with an equatorial diameter of 120,000 km (74,600 miles) (nearly nine-and-a-half times that of the Earth). It is particularly remarkable for its very low density of only 706 kg per m³, less than that of any other known planetary or satellite body, which indicates that, like Jupiter, it is primarily composed of hydrogen and helium. It, too, is expected to have a core of silicate materials, relatively larger than Jupiter's, surrounded by layers of metallic and molecular hydrogen. This core probably has a diameter of about 20,000 km (12,500 miles) surrounded by a 5000 km (3100 miles) thick layer of ices. The whole solid core probably contains about 10–15 Earth masses. Unlike Jupiter, the layer of metallic hydrogen is probably fairly small at about 8000 km (5000 miles) thick, and the major portion of the planet is formed of liquid molecular hydrogen, mixed with some helium.

Observations show that Saturn, like Jupiter, has an internal source of heat, and that the amount of energy is relatively more important, being about two to three times the radiation received from the Sun.

Saturn has an even greater polar flattening than Jupiter (about 10 per cent compared with 6 per cent) and the polar diameter is about 108,600 km (67,500 miles). The atmospheric rotation period was rather difficult to establish as the markings

visible from Earth were never so distinct as on Jupiter. The generally accepted period is $10_h14_m$ (System 1) with greater speeds towards the poles. Voyager experiments detected radio bursts at very long wavelengths which were shown to have a period of $10_h40_m$, this now being accepted as the sidereal period of the planet (System III).

As with other planets known to possess metallic cores and have rapid rotation rates (Jupiter and the Earth), a planetary magnetic field is generated in the interior. Before the Pioneer 11 and Voyager missions nothing certain was known about Saturn's magnetic field, but it has now been established that it is intermediate in strength between those of Jupiter and the Earth. It was something of a surprise when the geomagnetic axis was found almost to coincide with the rotational axis – they deviate by only 0·7 degree – as one theory for the generation of planetary magnetism required a much greater divergence. Just as Io is the primary source of the charged particles and atoms within the Jovian magnetosphere (with perhaps some contribution from Europa), so Titan provides most of the material in Saturn's system, with some gas also possibly derived from the inner satellites Enceladus, Tethys and Dione.

Titan orbits near the position of the magnetopause, although apparently it is usually inside it. It contributes large amounts of nitrogen from its atmosphere, as well as methane, which is broken down and provides the basic source for the large hydrogen torus found to encompass the orbits of Titan and Rhea.

All the satellites within the radiation belt affect the population of charged particles at their respective distances, but Titan in particular causes strong magnetic and plasma effects behind it. The number of charged particles drops dramatically at the edge of the rings, which essentially sweep up all particles within that radius.

The great glory of Saturn, though, is its ring system. There can be few of us who have not gazed in wonder at photographs of the wonderfully elegant disc that encircles the planet. Yet there is literally almost nothing to the rings: when seen edge on, they are so thin that they are almost invisible, and cannot be more than a few kilometres

### Saturn and Earth compared

|  | Saturn | Earth |
|---|---|---|
| equatorial diameter (km) | 120 660 | 12 756 |
| sidereal period of axial rotation | 10h 40m | 23h 56m 04s |
| inclination to orbit | 26° 44′ | 23° 27′ |
| density (kg per m³) | 706 | 5 517 |
| mass (Earth = 1) | 743·6 | 1·0000 |
| surface gravity (Earth = 1) | 1·159 | 1·0000 |
| escape velocity (km per s) | 36·26 | 11·2 |
| albedo | 0·76 | 0·36 |
| mean Sun-Saturn distance | 9·5388437 AU | |

thick. Although they look solid, they are very definitely not. In fact, from time to time, stars can be seen through some of the thinner parts of the rings.

It is clear that the rings contain only a tiny fraction of the total mass of Saturn, and that they must consist of myriads of tiny particles orbiting around the planet. It seems most likely that the particles are small chunks of ice, perhaps with small quantities of rocky material, but how and when they came to form such an elegant system of rings remains to be clarified.

Hopefully, we will learn more about the rings when the next space probe to visit the Saturn system, the US–European Cassini mission, is launched. The mission plan calls for a launch aboard a Titan IV/Centaur rocket in 1996 and a Saturn rendezvous six years later in the year 2002. Cassini will be similar, in some respects, to the Galileo mission to Jupiter, but its computers and imaging systems will be more advanced.

Once it has entered a highly elliptical orbit about Saturn, Cassini will begin a complex series of manoeuvres that will enable it to release a descent probe into the atmosphere of Titan which is Saturn's largest moon. The spin-stabilized probe will return a stream of data on the conditions in and the structure of the atmosphere for hopefully up to three hours.

*Above*: A composite Voyager picture (in true colour) of Saturn also showing three of the satellites – Tethys, Dione and Rhea. The black spot on the southern hemisphere of the planet is the shadow of Tethys. Some of the 'spokes' can be seen (particularly left) on the bright B Ring.

*Left*: A close-up view in false colour of two of Saturn's rings – the C Ring (blue in this picture) and the B Ring (orange). Seen from this close, the two rings are clearly made up of many smaller ringlets, of which more than 60 are evident in this frame.

## Uranus and Neptune

Although small in comparison with Jupiter and Saturn, the next two planets are still large bodies, and the four planets are sometimes known as the "gas giants". Uranus and Neptune are very similar in size (equatorial diameters of 50,800 km (31,500 miles) and 48,600 km (30,200 miles) respectively) and mass, with Neptune being about 15 per cent more massive than Uranus. Their characteristics are so similar that they may be conveniently discussed together. The densities of both planets are low and they probably have identical internal structures, with rocky cores surrounded by layers of ice and molecular hydrogen. Unlike Jupiter and Saturn, Uranus has a temperature that agrees with that calculated for its distance, indicating that it has no major internal source of heat. Neptune, on the other hand, does have a slight heat excess. Voyager 2 provided considerable information about Uranus from its fly-past in January 1986 and should do the same for Neptune in 1989.

Uranus is unique among the planets in that its equatorial plane (as shown by the orbits of the satellites) is almost perpendicular to the orbital plane. The axis is tilted by an angle of about 98°, so that, strictly speaking, the rotation and the orbits of the satellites are retrograde. The cause of this is unknown, but it has been suggested that the impact of a very large body could have been responsible. The effect of the axial inclination is that for very long periods of time (Uranus has an orbital period of about 84 years) the poles face towards the Sun, and calculations indicate that over an orbital period they receive more solar radiation than the equator. Voyager 2 observations indicate, however, that there is little temperature difference between the poles and equator of Uranus, indicating that an efficient heat-transfer process is at work.

Neptune's magnetic field remains unknown, but Uranus has a strong field (about 50 times the strength of the Earth's). Amazingly, this is inclined at 60° to the rotational axis and offset by about 8000 km (5000 miles) from the planet's centre. Considerable changes arise in the magnetosphere as the planet rotates, and radio observations of these variations show that the basic rotational period is $17_h14_m$. In visible light Uranus is practically featureless, nearly all detail being hidden by a high-altitude hydrocarbon haze, but Voyager did detect a few faint cloud features at middle and high latitudes. These rotated round the planet with periods between $16_h$ and $16_h54_m$, showing the existence of high winds – approximately 700 km/h at 60° latitude – in the same direction as the rotation. These winds are stronger at greater heights and are probably matched by flow in the opposite direction in the equatorial regions.

In the infrared Neptune shows some vague cloud features, with rotation periods ranging between $17_h42_m$ and $19_h36_m$, suggesting that wind speeds there may also vary with latitude. There are striking changes in the brightness of the northern and southern hemispheres over a period of years, indicating major alterations in the cloud cover, which probably consists of particles of frozen methane.

Uranus shows some auroral activity on its night side, but any present on the day side is swamped by an unusual, ultraviolet "electro-glow", caused by the collision of excited electrons with hydrogen molecules high above the planet at a height of about 1500 km (930 miles) above the cloud tops.

The rings around Uranus were initially discovered through observations of a stellar occultation in 1977. There are now known to be 10 rings, lying between about 42,000 km (26,100 miles) and 51,000 km (32,000 miles) from the centre of the planet. They are not all circular, nor do they all lie in the equatorial plane. Unlike Saturn's rings, they are very narrow, with widths ranging between 2 and 100 km (62 miles). These widths are not constant: that of the outermost, epsilon, ring varying between 20 and 96 km (60 miles), for example. The density of ring particles also varies, being greatest where the rings are narrowest, as might be expected. The particles are very dark (albedo about 0·05), so probably consist of some carbonaceous material, rather than of ice. Their size-range covers a few cm to a few metres. (This is in complete contrast to the situation at Saturn, where the rings are not only formed mainly of ice, but contain large quantities of much smaller particles.)

In addition to the rings, Voyager discovered several partial ring arcs, and about 100 bands of microscopic dust, with particles about 0·02 mm in size, spread throughout the ring system. Surprisingly, the main rings themselves are relatively free of dust.

### Uranus, Neptune and Earth compared

|  | Uranus | Neptune | Earth |
| --- | --- | --- | --- |
| equatorial diameter (km) | 50 800 | 48 600 | 12 756 |
| sidereal period of axial rotation | $17_h 14_m \pm 1_m$ | $17·5_h$? | 23h 56m 04s |
| inclination to orbit | 97° 53′ | 28° 48′ | 23° 27′ |
| density (kg per m³) | 1 270 | 1 700 | 5 517 |
| mass (Earth = 1) | 14·6 | 17·2 | 1·0000 |
| surface gravity (Earth = 1) | 1·11 | 1·21 | 1·0000 |
| escape velocity (km per s) | 22·5 | 23·9 | 11·2 |
| albedo | 0·93 | 0·84 | 0·36 |
| mean distance from Sun | 19·181843 AU | 30·057984 AU | |

The partial arcs may originate in the collision of small satellites, the debris gradually spreading out to form very tenuous rings. Micrometeorite bombardment of such faint rings probably gives rise to the bands of dust, which have very short lifetimes, and rapidly decay into the upper atmosphere of Uranus.

The discovery of rings around Jupiter, Saturn and Uranus has intensified the search for comparable features around Neptune. As yet there is only rather tenuous evidence (from an occultation) that a partial ring arc may exist rather close to the planet, between 28,500 and 32,500 km from its centre.

Uranus has fifteen satellites, while Neptune is known to have two or possibly three, one of which, Triton, is exceptionally large. The ten small satellites of Uranus discovered by Voyager 2 include two (named Cordelia and Ophelia) that act as "shepherds" to the outermost (epsilon) ring, while the other eight orbit between the rings and Miranda. All are very dark, like the rings, and most probably consist of carbonaceous material.

Of Uranus's large satellites, Oberon is saturated with craters, indicating an ancient surface, with just a few faults and some craters with dark (presumably flooded) floors. Titania has many signs of tectonic activity such as ravines and faults, but has few craters, an indication that major resurfacing has taken place. Umbriel is remarkably dark and relatively featureless, with just two bright features, one a distinct bright ring. The satellite's appearance is as difficult to explain as that of Iapetus and it may be covered by a thick blanket of dark material, which has only been breached in two places. Ariel, by contrast, is much brighter and shows evidence of even greater geological activity than Titania. It has an extensive network of rifted valleys, with flow patterns that suggest that a mixture of ice and rock has welled up from fractures in the valley floors.

Miranda has probably the strangest surface of any satellite in the Solar System. Part consists of densely cratered plains very typical of old surfaces. Within the plains, however, are strange areas known as 'ovoids' (or *circi maximi*), variegated regions surrounded by bands of parallel grooves and ridges. There are also great fracture zones, where some cliffs are as much as 10 to 20 km (6–12 miles) high. The ovoids are difficult to explain but appear to have arisen from partial differentiation of the rocky and icy masses that formed the satellite.

Oberon and Titania appear to contain more rock than Iapetus and Rhea, similar-sized bodies in Saturn's system. This might help to explain their greater geological activity as radioactive elements in rock would contribute to their internal heating. Similar considerations may apply to Ariel and Miranda, and the other satellites in the outer Solar System that appear to have been geologically active. Some heating may have arisen from short-lived isotopes such as iodine-129 and aluminium-26, which could have been produced in a nearby supernova explosion.

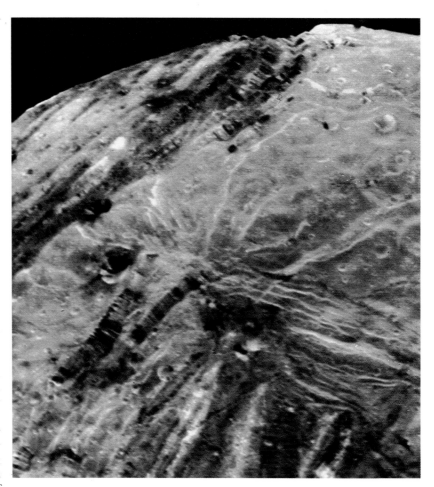

The surface of Miranda is quite unlike that of any other body in the Solar System. Its geology is exceptionally complicated and very difficult to explain.

Neptune's large satellite Triton poses problems of a dynamical nature. Its highly inclined retrograde orbit, which is very close to the planet, is very hard to explain if the satellite formed at the same period as the planet, when it should have accreted in the equatorial plane and orbited in the same sense as the planet's rotation. If it had been captured later, it would be expected to have a fairly eccentric elliptical orbit, rather than a perfectly circular one. Recent observations show that Triton's orbit is decaying, and that the satellite is likely to be disrupted by Neptune's gravitational forces in some 10 to 100 million years.

It used to be suggested that Pluto was an escaped satellite of Neptune and that this could have affected Triton's orbit, but this is now considered to be unlikely, especially since the discovery that Pluto has a large satellite. There is evidence that a methane atmosphere exists on Triton, and some recent results suggest that liquid nitrogen and methane "seas" may occur on the surface, together with more complex compounds. Confirmation of these theories will have to await the Voyager 2 fly-by.

Nereid, by contrast, has a very distinct orbit, with a higher eccentricity than that of any other known satellite. It possibly resembles other small icy satellites found in the outer Solar System. There is also some evidence suggesting that Neptune may have a third satellite, perhaps 180 km (110 miles) across and orbiting about 50,000 km (31,000 miles) from the planet.

## Pluto

The mean distance of Pluto from the Sun is so great (39·4 AU) that it is very difficult to study. It has not proved an easy task to measure the size of this planet, but it may be estimated from observations that the planet is covered in methane frost. Such a surface layer would have an albedo of 40–60 per cent, which, from the known brightness, leads to a diameter of about 3000 km (1900 miles) if it is completely frost-covered.

In the past, only very rough estimates of the density and mass have been possible. Despite the fact that Pluto was discovered in 1930 close to the predicted position calculated from the assumption that an unknown planet was disturbing the orbit of Uranus, these calculations now seem to have been erroneous, and Pluto apparently has little effect on Uranus and Neptune. In 1978, it was found that Pluto has a moon of its own, this was subsequently named Charon. The total mass of the system seems to be only about one-fifth of that of the Moon alone.

The diameter of Pluto appears to be close to 2300 km (1400 miles) while Charon could be as large as 1300 km (310 miles) and probably orbits at a distance of approximately 20,000 km (12,000 miles). The density of both bodies is very low. The orbital period of Charon seems to be of the order of 6·39 days. Charon has a highly inclined orbit, and could well be orbiting in the plane of the planet's equator.

Although it had been long suspected that Pluto had no atmosphere, and that the surface temperature would lead to any gases being permanently frozen, there is now some evidence that gaseous methane does exist, but the temperature is about 60 K, which is some 213° colder than water ice.

The planet's orbital inclination and eccentricity are far greater than those of any other planet, including Mercury, and for part of its orbit the distance from the Sun is less than that of Neptune. In fact, since 1978 Pluto has lost the honour of being the most distant planet.

With confirmation that Pluto is of low mass, the problem of the motions of Uranus and Neptune remains unresolved. It could be that there is a body, as yet undiscovered, with sufficient mass to cause the observed orbital disturbancies. The suggestions as to what this could be are: another planet, a dark stellar companion to the solar system and, inevitably, a black hole! A planet would have to be quite large and massive, and even if it were in a highly inclined orbit would have been unlikely to escape detection. The most probable candidate body is another star, perhaps a "brown dwarf". Discovering such a body, which could be located at quite a distance from the Sun, is likely to be very difficult, although it may possibly be feasible to detect any gravitational influence on the two Pioneer spacecraft which are now heading out into interstellar space in approximately opposite directions. For as long as they continue to operate it will be possible to determine from the Doppler shift of their radio signals whether they are being subjected to any additional gravitational influences.

### Pluto

| | |
|---|---|
| equatorial diameter (km) | 2290 |
| sidereal period of axial rotation | 6·39 |
| inclination to orbit | 118? |
| density (kg per m³) | 1 000? |
| mass (Earth = 1) | 0·0025? |
| surface gravity (Earth = 1) | 0·03? |
| escape velocity (km per s) | 1? |
| albedo | 0·40–0·60? |
| mean Sun-Pluto distance | 39·4 AU |

## Meteorites

Although the arrival of a meteorite may be a spectacular event, accompanied by a blazing fireball, the objects which reach the ground have little superficial appeal – most are rather drab chunks of stone, indistinguishable to the untrained eye from any other chunk of stone. To students of the Solar System, however, meteorites have an interest and importance which is out of all proportion to their abundance and dull appearance, because with the exception of the samples returned from the Moon, they provide the only pieces of the Solar System (besides the Earth itself) that can be studied directly.

Although shooting stars are common, it is only rarely that solid fragments survive their headlong passage through the Earth's atmosphere. Only a handful of fist-sized chunks are retrieved each year, and the total number of known meteorites is only a few thousand. Very occasionally, large meteorites arrive out of the blue – a 2,000 kg. (4,400 lb.) meteorite fell near Pueblito de Allende in Mexico in 1969, and provided a bonanza for meteorite students.

Clearly, a large meteorite could cause considerable damage if it fell in a built-up area, but the fact that there are no recorded deaths due to meteorite impacts anywhere in the world indicates just how rare they are. Many extremely

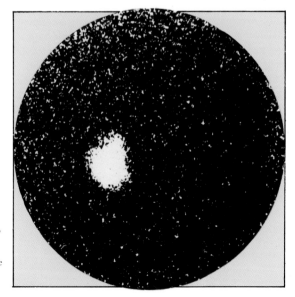

A photograph showing the combined images of Pluto and its satellite which orbits from north (top) to south. The orbital plane is highly inclined to the plane of the sky. If the satellite has an albedo similar to Pluto's, it may have a diameter as great as 1300 km (800 miles) and be the largest satellite in the Solar System relative to its primary.

large meteorites which fell in prehistoric times are known. Of these the largest is the Hoba meteorite, which fell in South Africa, and had a mass of at least 60 tonnes. Still larger meteorites are known not from their own remains, but from the huge craters that they blasted out on impact. Best preserved of these is Meteor Crater in Arizona, believed to have been excavated some 20,000 years ago.

Of all the known meteorites, those known as 'stones' are much the most common. They are composed dominantly of silicate minerals like those in the Earth's mantle. Much better known, because they are so much more distinctive are the 'irons'; these constitute only a small proportion of meteorites observed to fall, but a much larger proportion of them are found. They are composed of iron and nickel· alloys, sometimes associated with silicate minerals. It's thought that the Earth's dense core is probably composed of very similar materials; indeed it was the existence of iron nickel meteorites which gave support to this concept of the core before sophisticated geophysical techniques became available.

Drabbest of all meteorites, but of greatest importance, are the so called 'carbonaceous chondrites'. These dark coloured lumps of stone contain, as their name implies, substantial quantities of carbon. They are important because it is thought that they have a composition very close to that of the Sun, minus the Sun's light gases.

*All* meteorites are very ancient, around 4,600 million years old, the same as the oldest rocks recovered from the· Moon, and it is confidently supposed that the age of meteorites is also the age of the entire Solar System. Thus meteorites are very ancient, primitive components of the Solar System. But whereas the stony and iron meteorites do show evidence of some processes affecting them after the initial formation of the Solar System, the carbonaceous chondrites do not: they are the most primitive material sampled, containing clues to the formation of the Solar System some 5,000 million years ago.

## Asteroids

The asteroids are amongst the most poorly known members of the Solar System, which is scarcely surprising in view of the fact that they are perceptible only as tiny points of light in the most powerful telescope. The largest asteroid, Ceres, was also the first to be discovered, on the first day of the 19th century. Ceres is about 1,000 km. (625 miles) in diameter. Since its discovery, smaller asteroids have been found at an ever increasing rate. It is now thought that there may be as many as 70,000 of these objects, most of them less than 100 km. (62.5 miles) in diameter. The larger ones are more or less conventional, spherical bodies, like mini planets, but most of the rest are irregular chunks of rock.

Finding out what asteroids are made of is clearly a daunting task, given their small size and great distance from the Earth. Over the last few years, however, it has become clear that there

are strong similarities between the major classes of meteorites and asteroids. Thus asteroids of stony, iron and carbonaceous chondrite types have all been recognized. This was a great step forward and helps to explain the origin of at least some of the meteorites that reach Earth. Collisions taking place in the crowded traffic lanes of the asteroid belt may cause some fragments resulting from the impact to be flung off in orbits that intersect the Earth's. There are even a few asteroids known to have similar orbits. It may be that the only difference between meteorites and asteroids is one of scale.

At first sight, it seems the Earth has somehow escaped battering that has cratered all the other inner planets. On the contrary, Earth has received its fair share of impacts in the remote past. The effects, however, have been largely eroded away. The Solar System is now almost depleted of large-scale wandering interplanetary debris, but some rocks have fallen in recent times, as shown by the meteorite that struck Mexico in 1903 (*top*), and the existence of the huge Meteor Crater in Arizona, the result of an impact some 20,000 years ago.

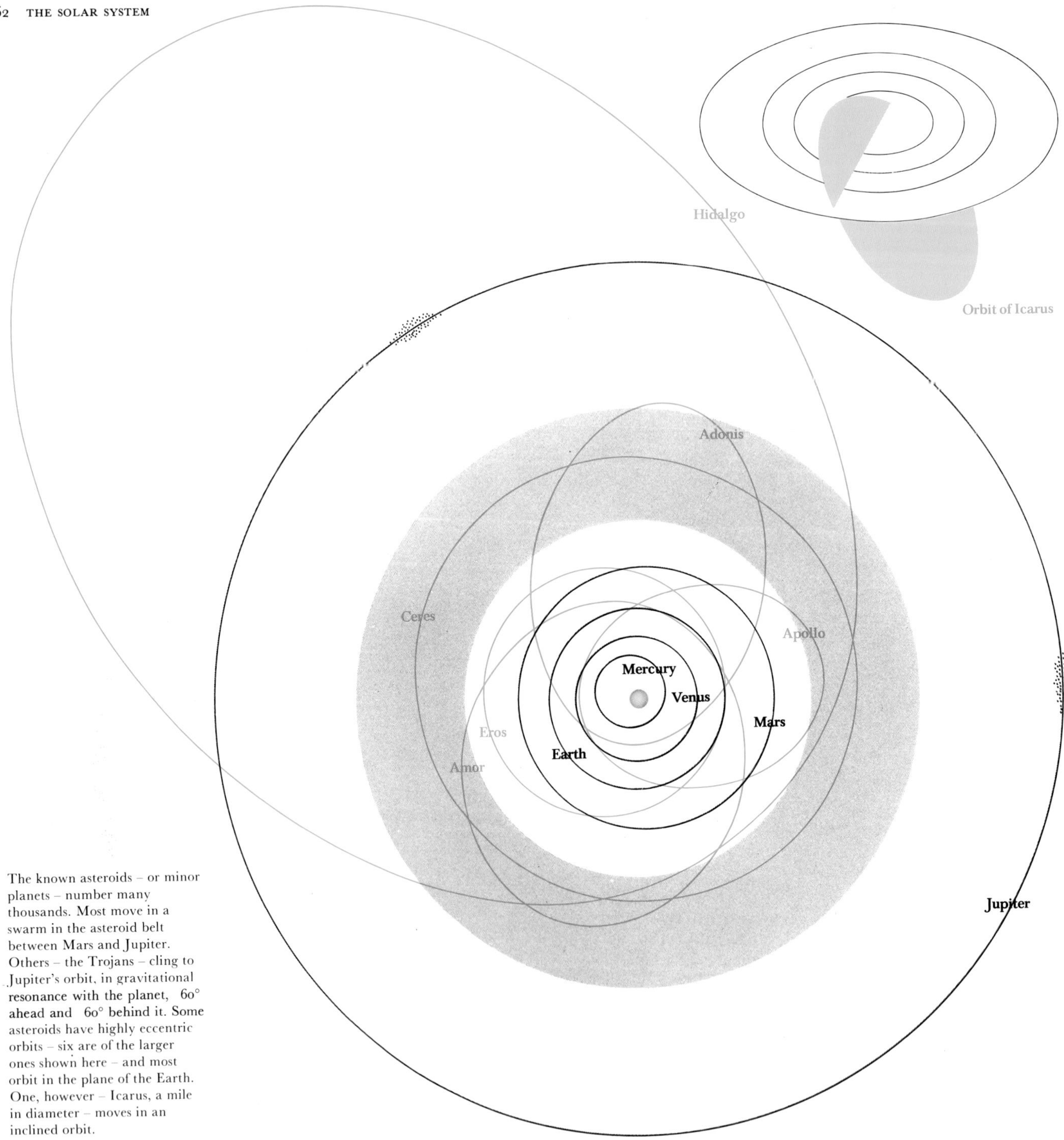

Hidalgo

Orbit of Icarus

Adonis

Ceres

Apollo

Mercury

Venus

Mars

Eros

Earth

Amor

Jupiter

The known asteroids – or minor planets – number many thousands. Most move in a swarm in the asteroid belt between Mars and Jupiter. Others – the Trojans – cling to Jupiter's orbit, in gravitational resonance with the planet, 60° ahead and 60° behind it. Some asteroids have highly eccentric orbits – six are of the larger ones shown here – and most orbit in the plane of the Earth. One, however – Icarus, a mile in diameter – moves in an inclined orbit.

## Comets

Although comets which are visible to the naked eye are not common, it is thought that out beyond the orbit of Pluto myriads of comets orbit the Sun in complete anonymity, never coming near enough to be visible. It has even been postulated that there is a great cloud of them, known as the Oort cloud, surrounding the visible parts of the Solar System.

The appearance of a great comet is undoubtedly one of the most striking celestial phenomena which can be seen by the naked eye. The fact that the occurrence of comets is generally unpredictable, and that they may suddenly appear as large and prominent objects, only serves to make them the more remarkable. However, it is now known that vast numbers are so faint that they can only be

observed with the largest telescopes and that many others must be missed because, as viewed from the Earth, their orbits are too close to the Sun for them to be seen. Study of their frequency and their orbits suggests that comets in the Solar System number many millions.

Like the minor planets, so many comets are known that a special classification has been introduced. It is usual for the name of the discoverer (or discoverers, up to a maximum of three) to be given to a comet, although this is sometimes varied, as in the cases of Comets Halley, Encke and Crommelin, by using the name of the person who has made extensive orbital calculations. Further identification is given by the year and a letter awarded in order of discovery. If the comet is found to be periodic the letter P is used as a prefix. At a later date, a final designation is given in accordance with the date at which the various comets came to perihelion by giving the year in which this occurred and a Roman numeral. For example Comet Bennett 1969i (the ninth to be found in 1969) became Comet Bennett 1970 II (the second to pass perihelion in that year). Recoveries of periodic comets are included in this scheme and, occasionally, when a comet has been lost for a considerable time, the name of the rediscoverer is added to that of the original finder, as with Comet P/Perrine which was not recovered for five periods, and was then found by Mrkos. The object is now known as Comet P/Perrine-Mrkos.

The vast majority of comets which have been fully studied have closed (elliptical) orbits and are thus true members of the Solar System. Some orbits are so eccentric that, initially, for computational purposes, they may be treated as *parabolic*. Calculation shows that those comets with *open hyperbolic* paths which are escaping from the Sun's influence have been perturbed by the planets (especially Jupiter). Despite the fact that comets have been lost by the Solar System, none have been observed to enter it from interstellar space along hyperbolic paths. There are indications that some comets have aphelia at many thousands of astronomical units, perhaps even halfway to the nearer stars, but accurate determination of aphelion distances and periods in such cases is extremely difficult. It has been suggested that a large reservoir of comets (the Oort Cloud) exists at such great distances, and that occasionally they are perturbed by the gravitational effects of nearby stars into orbits that enter the Solar System.

A distinction may be drawn between cometary periods which are of thousands or even millions of years and those which are comparable with the planets. The division is somewhat arbitrary, but is generally taken as being at about 200 years.

At great distances from the Sun, comets are small, faint, indistinct objects, but as they approach, volatile materials begin to be vaporized to produce the main head or coma. At times this may become exceptionally large; in the Great Comet of 1811, for example, the diameter exceeded $2 \times 10^6$ km (nearly one-and-a-half times that of the Sun). The tail may begin to develop at a considerable distance, as happened with Comet Schuster 1976c, which has the greatest perihelion distance known (6·882 AU, that is, beyond Jupiter) but which nevertheless had a moderate tail. Tails may not only be highly conspicuous, but also exceptionally long; that of the Great Comet of 1843 (a Sun grazer) had a length of about $3 \cdot 2 \times 10^8$ km, considerably greater than the mean distance of Mars from the Sun. At times, a small point-like nucleus, may be seen in the centre of the coma. Occasionally, multiple nuclei develop. A feature which can only be observed from spacecraft, is a vast hydrogen halo surrounding the visible portions of the comet.

Various spacecraft investigated Comet Halley in 1986, particularly the Soviet Vega 1 and 2 probes,

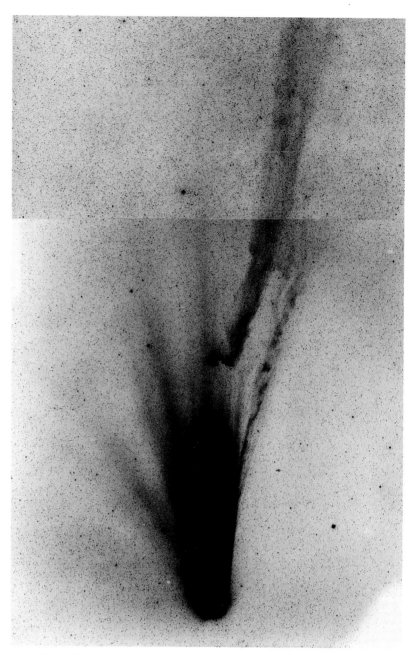

and ESA's Giotto. They provided one unexpected result: the density was remarkably low, perhaps one-quarter of that of ice. The total mass is only around $6 \times 10^{13}$ kg (less than one-thousand-millionth of that of the Moon). Despite the enormous apparent size of the tail and coma, the nucleus itself must be mainly empty space, and it has been graphically described as a "dirty snowdrift", consisting of a loose aggregate of ice and dust particles, that must have formed under very cold conditions at a great distance from the Sun. The nucleus is irregular, being roughly 15 by 8 km in size, and has an exceptionally low albedo (about 0·05), indicating that the surface is covered in a carbon-rich crust. The dark layer was breached in several places by major jets in which the gas and dust were released.

Comet Halley's ices apparently consisted of water, carbon dioxide and carbon monoxide, but there was little methane, contrary to expectations. Other elements found were carbon,

This negative image of Comet Halley is a composite of two photographs taken on March 10 1986. The relatively featureless dust tails curve away to the left (North), while the much more complex gas tail extends towards top right. The distinct kink and break is known as a 'disconnection event'. This occurs when there is a reversal in the magnetic polarity of the solar wind sweeping past a comet, causing a temporary disruption of the ionized gas tail.

nitrogen, oxygen, silicon, sulphur and magnesium – all at Solar System abundances. The dust consisted of three types: a silicate, similar to many rocks; a mixture of light and heavy elements, corresponding to no known mineral, and a carbon-rich compound of carbon, hydrogen, oxygen and nitrogen. The ices give rise to one of the two components seen in cometary tails, the ionized gas tail, whose behaviour is controlled by the solar wind. The solid particles form the generally featureless dust tail.

Cometary particles are known to be responsible for many meteor showers, but the density of a cometary tail is so low that passage of the Earth through it is not likely to produce any observable effects. On very rare occasions, collision with the main body of a comet may be expected, and this is almost certainly the explanation for the brilliant fireball and immense explosion which occurred on 30 June 1908 in the Tunguska River area of Siberia. Trees were uprooted as far away as 40 km, and the pressure waves recorded as far off as the British Isles. Yet no major fragments have ever been found, only microscopic iron and silicate particles having been recovered from the soil. These results are consistent with an encounter with the head of a small comet largely composed of ice and small solid particles.

## The origin of the Solar System

In examining the origins of the Solar System we take on one of the largest and most difficult problems in Science. Not only are we dealing with an event that took place over 4,600 million years ago, but are also limited to deductions drawn from telescopic observations, data from a score or so spacecraft, a small quantity of materials from the Moon, and a random selection of meteorites that happen to have reached the Earth.

Any hypothesis for the origin of the Solar System has to be able to account for a number of different factors, and this helps to constrain the possible models:

- the first factor concerns the distribution of mass within the Solar System. The innermost planets are small, dense and rocky, the outer ones large, of low density, and made largely of volatile materials;
- second, and not immediately obvious, is the rate at which different members of the Solar System rotate on their axes. Jupiter, for example, spins much faster than the Sun, which is the opposite of what one might expect from first principles;
- third, is the bulk composition and internal structure of the planets;
- fourth is the actual evidence of early events in the history of the Solar System contained on the surface of the inner planets;
- fifth is the need to explain the existence and nature of the swarms of meteorites, asteroids and comets and their relationship to the major planets and to each other;

– and finally one has to take note of the very subtle chemical clues contained within the most ancient material at our disposal, in meteorites.

Carbonaceous chondrite meteorites such as the one that fell at Pueblito de Allende in 1969 have proved to contain a treasury of data on the extremely early history of the Solar System. The evidence is not obvious, but is locked up on complex ratios of elemental isotopes. Parts of the Allende meteorite were found to be extraordinarily rich in an isotope of magnesium known as $^{26}Mg$. Now this isotope can only be formed by the radioactive decay of an isotope of aluminium, an isotope whose half life is so short that it no longer exists anywhere in the Solar System. Physicists have concluded that the only way that this aluminium isotope could have been produced was in the catastrophic explosion of a nearby supernova, and it is this event which they presume to have provided the trigger to the formation of the Solar System.

The story is even more complex than this, however, because other radiogenic isotopes in meteorites indicate that there may have been another, slightly earlier supernova explosion as well, which seeded the meteorites with quite different isotopic species. Both these stupendous explosions seem to have taken place a couple of hundred million years before the actual formation of the Solar System 4,600 million years ago.

The isotopic data is the last firm evidence that we have to go on. The rest is speculation. The present consensus is that the supernova explosions sent great pressure waves through a diffuse cloud of dust and gas that constituted the primitive solar nebula, the site of the formation of the Solar System. Perhaps there was already a local accumulation of gas beginning to glow under the heat of its own collapse – a proto-Sun. The pressure waves may have initiated the condensation of the cloud of dust and gas into a more confined space, where individual particles could begin to interact with one another gravitationally. The explosions also sprayed unmistakable radiogenic isotopes throughout the nebula.

But what were the constituents of this great cloud of dust and gas? It seems certain that most of the cloud was made up of hydrogen gas, the most abundant element in the universe. The solid particles were probably ice, and frozen gases, with small quantities of rocky and metallic material. They may have resembled tiny, fluffy, rather dirty snowflakes.

With the passage of time, these small particles interacted with one another, and began to clump together to form larger and larger bodies, perhaps several hundred metres in diameter. At this stage, the bodies were homogeneous all through: they were simply large, dirty snowballs in space. The analogy with comets is obvious – the comets we see today may be merely relics of this early stage of the Solar System.

Subsequently, things became more complicated. For one thing, as the proto-planets continued to accrete together, their masses became such that impacts between them began to destroy some of them, breaking them down once again into smaller fragments. For another, the Sun now began to exert its influence, and this soon became the most important single feature in explaining the differences between the planets. Those nearest the Sun accreted under high temperatures; those furthest from the Sun under progressively decreasing temperatures. Thus the proto-planet that was to become Mercury was quite unable to accrete volatile materials such as water, gases and even low melting point metals. Uranus, by contrast, was formed in a very chilly part of the Solar System, and was thus able to incorporate into itself even the most highly volatile elements. There is therefore an obvious increase in the volatile content of the planets outward from the Sun, a progression which is slightly complicated by the fact that Jupiter was so massive that the effects of temperature on it were relatively unimportant.

Thus, the accretion of the planets from discrete grains of dust into solid, individual entities was tightly controlled by temperature. Once the planets had formed, however, a whole series of further changes took place within them.

The process of accretion itself liberated large amounts of energy, which caused the inner planets at least to heat up. The effects of this were compounded by heat liberated by short-lived radioactive isotopes trapped within the planets' interiors. The results of this heating are thought to have caused large scale, perhaps complete melting of the inner planets. The melting itself caused massive segregation processes to take place, with dense iron and nickel components drifting downwards to congregate at the centres of the planets, forming their cores. This process was accompanied by still further heating.

Core formation in the inner planets was probably completed within a few tens or hundreds of millions of years of their initial accretion, but the differentiation and segregation that took place above the cores continued long afterwards. It is these processes which led to the well defined threefold structure of the inner planets: core, mantle and crust.

The Earth's crust contains no record whatever of the accretion events; all evidence has long since been wiped away by continuing geological processes. On the Moon, Mercury, and to a lesser extent Mars, however, we can still see traces of those events. The process of accretion did not stop abruptly 4,600 million years ago. Large bodies continued to plough into the surface, throwing up enormous craters and ring basins on the Moon and Mercury, and they continued to arrive in large numbers until about 4,000 million years ago. After that, the number of bodies arriving seems to have dropped off steeply. The process of planet formation was essentially complete.

So much for the planets. But what of the asteroids and meteorites? As has been emphasised earlier, the carbonaceous chondrites are primitive bodies; little has happened to them. But the stony and iron meteorites have clearly experienced episodes of melting and differentiation. So, by analogy, have the asteroids. It seems that a large number of separate, small planetary bodies formed within the present asteroid belt, large enough for some internal segregation to take place, but not large enough for internal activity to persist more than briefly. Occasional impacts between these segregated bodies may have demolished some of them, so that fragments from both their metallic interiors and the stony exteriors occasionally arrive on Earth, 4,600 million years after their formation.

## Other planetary systems

Although it has been possible to gain a reasonable idea of the way in which the solar system was formed, it is difficult to obtain direct evidence for the existence of other planetary systems. However, there are various indications that such systems do exist and have been observed.

Strong evidence for protoplanetary systems has been provided by infrared observations, particularly by the IRAS spacecraft (Chapter 10), most notably for $\beta$ Pictoris, where the disk extends out to about 900 AU.

The minimum mass which can give rise to thermonuclear fusion and produce a true star is about 0·06 solar masses. Between this and about 0·01 solar masses objects will shine for about 1000 million years due to the release of gravitational energy as they contract. After this time they will become dark stars, sometimes called *brown dwarfs*. Any mass smaller than one hundredth that of the Sun may be described as a planet, as it will only radiate away heat at infrared wavelengths. This is the case for both Jupiter and Saturn.

Observations of Doppler shifts caused by the orbital motion of some stars have allowed astronomers to estimate the number of stars which have companions, and to assess how many of these companions might themselves be stars, brown dwarfs or true planets. It seems that the vast majority of stars have one or more companions. Although the calculations are, at present, rather uncertain, possibly about 20 per cent of solar type have planetary companions.

Photographic methods can chart the motions of nearby stars. In the absence of companions the track should appear as a straight line, but the presence of other bodies will cause a "wobble". This makes it possible to estimate the mass and distance of its companions. The work is very difficult, but at least seven stars are thought to have planet-sized companions. A different (spectroscopic) method appears to have found two probable systems, $\tau$ Ceti, and $\varepsilon$ Eridani – the latter already suggested by proper-motion work – and five "possibles". The Hubble Space Telescope (Chapter 10) may be able to detect directly large planetary bodies in orbit around nearby stars, but it is likely to be a very long time before Earth-sized planets are detectable.

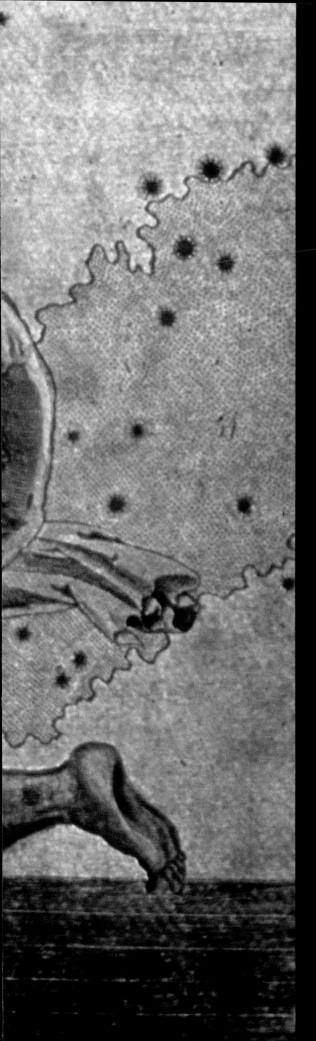

# 4/THE FRAMEWORK OF THE STARS

Stars look like mere pin-pricks of light, set close by in a dome of darkness. For centuries, men sought to understand the Universe within such a cosy framework, sketching pictures in the sky (like the 17th-century version of the constellation Auriga, the Charioteer, at left) to record the stars and provide a reassuring mythology. It seemed common sense that men's destinies and the cycles of the stars were intimately related – astronomy and astrology were different sides of the same coin. Only relatively recently – over the last three centuries – have we come to see that the stars are sun-like bodies set in the vastness of three-dimensional space. And only in this century have we been able to measure the distances between the stars and understand the nature of the Catherine-wheel structure – our Milky Way Galaxy – that binds the stars together.

This 11th-century Arabic version of the figure of Perseus points to the ancient origins of the names and figures associated with the constellations. Perseus was the Greek mythological hero who beheaded Medusa.

In this 14th-century view, two angels crank the fixed stars round the motionless Earth. The idea now seems naive, but (if the motion of the planets is ignored) it makes little difference observationally whether the Earth or the stars are considered to be in motion. The picture also reflects the belief that the wheeling heavens needed repeated inputs of divine energy to keep them in motion.

**I**t is not surprising that astronomy is the oldest science. While early man's interest in the changeable world around him was highly practical, concerned with the need to survive, his interest in the patterns of the stars and their regular cycles was of a different kind. His imagination could roam the stars unfettered. The sky seemed to roof the world; its exquisite pattern of stars called out for explanation. On this dark canvas, our ancestors depicted huge figures, outlined by scattered stars, much as children of today join up the dots in puzzle books to find the hidden figures. And as the stars bear no numbers to guide the inscribed lines, different cultures have seen different patterns in the same stars.

But even the esoteric field of star-watching had a practical relevance to ancient man. The rising and setting of stars during the night formed a night-clock for the Egyptians, complementing the shadow-stick sundial in use during the day. The appearance of different constellations at different seasons was a simple calendar. And to these early scientists, it was only logical that other appearances in the sky were related to terrestrial events. The Chinese kept a watch for temporarily bright 'guest stars' which surely must signal changes in the Empire; the Babylonians were more concerned with keeping track of the planets, the bright moving 'stars' which wander slowly among the constellations, and should also mirror changes on the Earth. To most early civilizations, there was no hard line between the subjects we would now call the science of astronomy and the occult lore of astrology.

The astrological interpretation of these portents in the sky could only be assessed by their positions relative to the constellation figures. To the Chinese, a guest star 'trespassed' against the constellation in which it appeared, and their astrological records of such happenings are an invaluable source of data on past astronomical events, a virtually complete record stretching back 2,000 years. The Moon and planets travel a well-defined track (see Chapter 4) through a band of constellations, known collectively as the zodiac, whose names have come down to us today to appear in the popular horoscope columns in newspapers and magazines. A glance at one of these will still reveal that important changes in our personal lives should be expected when a planet crosses the division from one zodiacal constellation to the next!

Astronomers now dismiss the claims of astrologers, but the ancient astrological signs have at least given both astronomers and the general public common patterns with which the heavens can be mapped.

The origins of the constellation patterns and names we use today are lost in the mists of the remote past, in the world of the ancient civilizations around the Mediterranean Sea. Our principal constellations were described by the Greek philosopher Aratus around 275 BC, but he was merely passing on an older tradition which probably sprang originally from Sumeria. The Greek astronomers Hipparchus and Ptolemy finalized a list of 48 constellations, which have come down to us in the form of Ptolemy's great work, *The Almagest*. These patterns, well-established throughout Western Europe in medieval times, were outlined by the brighter stars in the sky; between them were uninteresting regions populated by dim stars which were assigned to no particular constellation.

Some of the most striking of these ancient figures can be recognized without too much difficulty. (Ironically enough, the best place to learn the constellation patterns is from a city environment; with only the brightest stars visible, the shapes are much easier to spot.) The figure of Orion, striding through the January skies, is one of the most familiar. Bright stars mark his shoulders and knees, three central stars delineate his belt, while fainter stars make a hanging sword, a raised club and a lion-skin shield. Opposite him in the sky, and only poorly seen from northern latitudes, is the equally striking Scorpio, the scorpion. The bright star Antares marks his heart, and winding below is the body and tail, with a sting at the tip. In front, fainter stars which once represented his claws have now been lopped off as the separate constellation Libra, the scales.

To star-gazers in high northern latitudes, the seven stars of the Plough are an ever-present and familiar sight. They also act as a celestial signpost, for by following the lines of its stars we can readily find other constellations. In particular, the two 'Pointers' leading the Plough

The zodiac, the band along which the Sun, Moon and planets travel as seen from Earth, is shown in this mid-17th century painting. The Solar System is shown correctly with the Sun at its centre. The zodiac was first devised by Mesopotamian astronomers as a system of animal signs as early as 3,000 BC, and the circle of 12 constellations came to be called the '*zodiakos kyklos*' – 'circle of animals' – by the Greeks. Hence its present name, although the signs themselves are no longer all animals.

show the position of the Pole Star, a star which happens to lie almost directly above the Earth's north pole. The southern hemisphere's equivalent of the Plough is the Southern Cross, a superb little group of four bright stars, which can again be used as a pointer to other constellations.

To the amateur star-gazer, learning the constellations is the only way to become familiar with the night sky, and the bright patterns of antiquity are fairly easily recognized. The southern hemisphere astronomer, though, finds that the bright patterns near the south pole of the sky often bear strangely modern-sounding names. This region was invisible from the latitude of the Mediterranean civilizations, and the newly-discovered stars were ordered into 12 new constellations by Johann Bayer of Augsburg in 1603.

Once Bayer had broken with tradition, and boldly introduced his own figures into the sky, others were not slow to follow. Bayer's colleagues Julius Schiller produced a Christian star atlas in 1627, keeping the constellation shapes, but 'improving' the pagan names: Orion became St. Joseph, and the Great Bear the Ship of St. Peter. This innovation did not catch on, however, but other astronomers took the liberty of 'forming' constellations out of the faint stars lying between the classical constellations. Some of these have survived the test of time, while others have fallen into disuse. There is still a Unicorn near Orion, and a Giraffe not far from the Pole Star, and in the southern hemisphere some of the weird

creations of Abbé Nicholas de la Caille have passed into general use. These include a sculptor's workshop, a pendulum clock, an air pump, (perhaps more fittingly) a telescope and microscope, and last – and definitely least – Table Mountain (Mensa), which holds the distinction of being the faintest constellation in the sky. Mercifully, other constellations proposed by Johann Bode were soon dropped, otherwise our skies would also be cluttered with a log-line, an air-balloon, a printing press and an electrical machine!

Modern professional astronomers need to know positions far more accurately than simply by constellation: a position of the famous Crab Nebula given as 'above the southern horn of Taurus (the bull)' is of little use today. The sky, though still being treated as a sphere, has therefore been divided by imaginary lines, corresponding exactly to latitude and longitude on Earth. The latter run from the north to the south point of the celestial sphere (the points above the Earth's north and south poles), the 24 lines being spaced at 15° intervals. As the Earth rotates, these lines of *right ascension* pass overhead at the rate of one per hour: they are accordingly named as 'hours', and the intermediate spaces divided into 60 'minutes'. The equivalent of latitude is the astronomer's *declination*, measured in degrees north (+) or south (−) of the celestial equator towards the poles. So a position in the sky can be specified by two co-ordinates, just as

# Mapping the Stars

At first sight, understanding the way stars are pinpointed on a star map seems forbidding. The details are, indeed, complex; but the principles are not. The reality of three-dimensional space is ignored and the stars are seen as points on the inside of a sphere, which is mapped with a grid of lines equivalent to those of latitude and longitude (right). In this way the arbitrary patterns of the constellations (below) can be set in a fixed frame of reference.

The celestial equivalent of latitude – which measures distance north and south of the equator – is *declination*. The equator, which is set at $23\frac{1}{2}°$ to the plane of the orbit, is projected on to the 'celestial sphere'. A star's position north or south is then measured in degrees, between 0° at the equator and 90° at the poles ( + = north; – = south).

The second coordinate – the equivalent of longitude – is more problematical. As with longitude, a north–south baseline (like the Greenwich Meridian) must be fixed. The celestial meridian is defined by the point at which the path of the Sun crosses the celestial equator at the beginning of spring in the northern hemisphere, when day and night are of equal length – the vernal or spring equinox ('equal night'). From this point, a line is drawn north and south to the celestial poles to form the *prime meridian*.

Measurements can then be taken from the prime meridian on a grid drawn from the celestial equator. Since the stars' positions in the sky depend on the Earth's 24-hour period of rotation, the units of this coordinate are usually hours (although degrees are also used; 1 hr = 15°).

In practice, the system works as follows: An observer sees the points in the sky defined by the prime meridian swing above his head from east to west. The other stars follow hour by hour as the Earth turns. Thus the position of any star can be established by the amount of time it lags behind the meridian. This coordinate is known as *right ascension*. ('Right' has nothing to do with direction; it derives from an archaic use of 'right' meaning 'vertical' as in 'right angle'.)

But there are added difficulties. The Earth's axis is not quite stable. It wobbles, like an unsteady top, and the point of the axis *precesses* in a circle over 26,000 years. The equinoxes are therefore not at exactly the same time every year, and the celestial coordinates must therefore be given according to a particular year (at present, either 1950 or 2000).

Finally, when the system of coordinates was established 2,000 years ago by the Greeks, the spring equinox lay at a position in the constellation of Aries known as the First Point. Confusingly, the meridian is still known as the First Point of Aries, although precession has since brought it into Pisces.

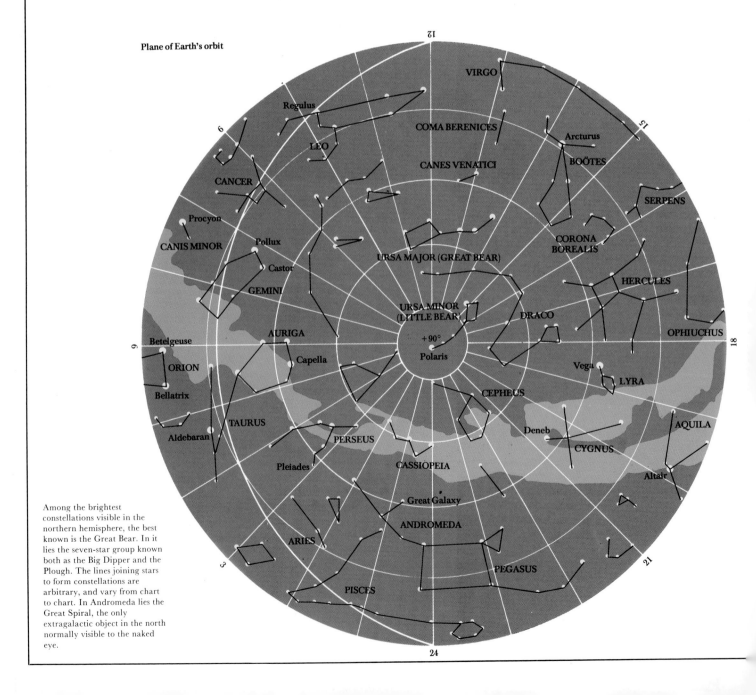

Plane of Earth's orbit

Among the brightest constellations visible in the northern hemisphere, the best known is the Great Bear. In it lies the seven-star group known both as the Big Dipper and the Plough. The lines joining stars to form constellations are arbitrary, and vary from chart to chart. In Andromeda lies the Great Spiral, the only extragalactic object in the north normally visible to the naked eye.

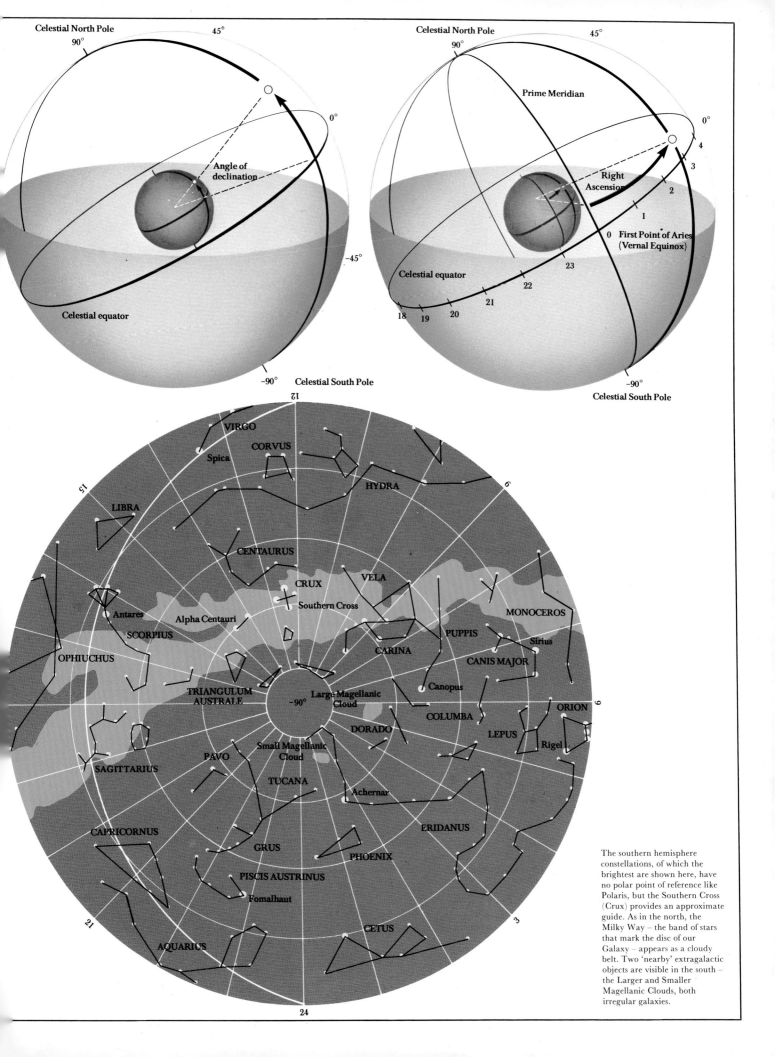

Celestial North Pole
90°
45°
0°
Angle of declination
Celestial equator
−45°
−90° Celestial South Pole

Celestial North Pole
90°
45°
Prime Meridian
0°
4
3
2
1
Right Ascension
0 First Point of Aries (Vernal Equinox)
23
22
21
20
19
18
Celestial equator
−90°
Celestial South Pole

−90°
Celestial South Pole

12
VIRGO
Spica
CORVUS
HYDRA
15
LIBRA
CENTAURUS
9
CRUX
VELA
Southern Cross
MONOCEROS
Antares
Alpha Centauri
PUPPIS
Sirius
SCORPIUS
CARINA
CANIS MAJOR
OPHIUCHUS
Canopus
TRIANGULUM AUSTRALE
−90°
Large Magellanic Cloud
COLUMBA
ORION
DORADO
LEPUS
Rigel
6
Small Magellanic Cloud
PAVO
SAGITTARIUS
TUCANA
Achernar
ERIDANUS
CAPRICORNUS
GRUS
PHOENIX
PISCIS AUSTRINUS
Fomalhaut
21
CETUS
3
AQUARIUS
24

The southern hemisphere constellations, of which the brightest are shown here, have no polar point of reference like Polaris, but the Southern Cross (Crux) provides an approximate guide. As in the north, the Milky Way – the band of stars that mark the disc of our Galaxy – appears as a cloudy belt. Two 'nearby' extragalactic objects are visible in the south – the Larger and Smaller Magellanic Clouds, both irregular galaxies.

This late Medieval sketch of the zodiacal sign Virgo shows her carrying in her right hand Spica – the 'ear of corn' star – which is still used as an alternative name for the constellation's brightest member, Alpha Virginis.

a place on Earth is pin-pointed by its latitude and longitude. The Crab Nebula, above the bull's horn, is better – if less poetically – described as having a right ascension of 5 hours 31 minutes and a declination of $+22°$.

Knowing the co-ordinates of an object, a professional astronomer can swing his telescope to it without any knowledge of the constellations. Indeed, it is rare to find a professional astronomer today with more than a fleeting acquaintance with the constellation patterns and names. In today's astronomy, too, instruments for detecting radio waves, X-rays and other radiations from space must be precisely aligned on the astronomical objects emitting them, and here the co-ordinate system is even more vital. After the position of a celestial X-ray source, for example, has been accurately measured, it can be compared with the positions of stars to identify the culprit 'X-ray star'.

The long tradition of astronomy is not altogether broken, however. In the early days of a new branch of astronomy, positions cannot be measured particularly accurately, and constellation names are often invoked to indicate a general region of sky. An X-ray source which is now believed to surround a black hole, one of the most controversial and exciting objects predicted by the new astronomy, is known by the name Cygnus X-1, the first X-ray source to be found in the constellation of the Swan.

Generally, however, modern astronomy has little use for the constellation patterns, for they have no existence in three-dimensional reality. The stars are not simply studded on the inside of a huge black vault, but are remote suns lying at different distances from us, and the patterns they appear to make are clearly fortuitous. Some of the stars making a constellation will be nearby, feeble stars; others are very distant superluminous beacons. With the departure of astrology as a non-science, astronomy has left behind the study of the constellations to move into the investigation of the real Universe, and the stars themselves are the jumping-off point for the investigation.

The brightest stars have always had individual names: some historians have argued that these stars were recognized before the constellations, and it is quite likely that the constellation Canis Major, the Great Dog, was named after its chief star, Sirius, the Dog Star. Most of our star names are Arabic, tacked on to Greek astronomical knowledge that passed through Arab lands during the Dark Ages, to reappear in Spain in the early Renaissance. So we find that the brightest star in the obviously classical constellation of Hercules bears the outlandish name Rasalgethi – 'Head of the Kneeler' in Arabic.

Johann Bayer, the constellation inventor, introduced the first systematic star names in his catalogue of 1603. In each constellation he distributed the letters of the Greek alphabet, generally in order of brightness. The most prominent star in the constellation Virgo is Spica – meaning 'ear of corn' – and in Bayer's

catalogue this star becomes Alpha Virginis. (Notice that despite the mixture of the Greek and Latin languages, classical grammar is strictly followed, and the constellation name appears in the genitive case. Incidently, this system also applies to constellations like the telescope and microscope, instruments unknown in classical times, whose concocted Latin names have invented genitive cases: Telescopium – Telescopii, for example.)

Bayer's Greek letters are still used for the brightest stars, apart from the score with well established names of their own. The nearest bright star, for example, is Alpha Centauri; its alternative name of Rigil Kent, used by air navigators, has never caught on in the astronomical world. But the Greek alphabet runs to only 24 letters – alpha to omega – and other star catalogues have grown up to name the multitude of fainter stars. When the position of the X-ray source Cygnus X-1 was pinned down precisely, for instance, it was found to coincide with a star already catalogued 40 years earlier, and bearing the number HDE 226868.

Measuring the positions of stars is a fundamental branch of astronomy, with its own name – astrometry – and a multitude of new instruments to refine the positions. But to those not engaged in astrometry, a range of other questions seems more pressing: How far away is the star? How luminous is it? How does it work?

So far, we have been thinking largely in the ancient star-gazer's terms, measuring positions 'on the celestial sphere', that great black bowl which seems to surround us at night. But the stars are not points of light: they are searingly hot, blindingly bright suns, much like our own Sun but reduced to gentle glow-worms as their light spreads out on the enormously long journey to us. To understand the Universe, we must fill in the third dimension. Knowing the stars' distances, we can find out how they are arranged in space around us, and we can also calculate their intrinsic properties. Then the full battery of modern physical theories can be brought to bear on the stars, to force from them the secrets of their inner workings, and of their birth and death.

## Adding the third dimension

At first sight, measuring the distances of the stars seems to be a stab at the impossible. Even in the most powerful telescopes, few stars appear more than unresolved points. One point of light in the sky is so much like any other that it is difficult to see how an astronomer can confidently assert that one is three, 10 or 100 times further away than another – and give definite distances for them. And when we realize just how large these distances are in everyday terms, the problem seems even more intractable.

The Sun, our local star, is 150 million km. (93 million miles) from the Earth. This distance is huge compared with our everyday experience: in a supersonic airliner a one-way trip to the Sun would take all of ten years! Yet when we venture into the stellar universe, this is a microscopic distance. If we took the Sun away so far that its light dwindled to that of a star like Sirius, we would have moved it 100,000 times further away – a jet journey of a million years. We are now dealing with distances on the grand scale – even a million million kilometres is too small a unit to use conveniently. Instead, astronomers resort to a unit which happens to be about the right size, the distance which light travels in one year. The speed of light is unimaginably high – at 300,000 km. (186,000 miles) per second, a light beam would travel round the Earth seven times in less than one second. Over the course of a year, light travels almost 10 million million kilometres – and this distance, one *light-year* is the fundamental unit we shall use for measuring stellar space.

In terms of light-years, the nearby stars have cosily small distances. Alpha Centauri is 4.3 light-years away; its faint companion Proxima is very slightly nearer and so makes the grade as the nearest star to the Sun. Other well-known stars have distances of a few dozen, or hundreds, of light-years, out to about 900 light-years for Rigel, the bright bluish-white star in Orion.

These distances are daunting on the human scale, but they can be plumbed by astronomers. There are in fact several ways of measuring star distances – some applying only to particular kinds of objects like double stars or variable stars – but one stands out above all others: the method of parallax. The basic principle is fairly familiar. If you hold a finger in front of your eyes, and then open and close your eyes alternately, the finger seems to move to and fro against the

Our nearest stellar neighbours are seen here set in a cube with sides 12 light years in length. Many of them, like the close-by Barnard's Star, are too dim to be visible to the naked eye. Some of the brighter objects – Alpha Centauri, Sirius, 61 Cygni – are in fact double stars whose members are impossible to separate on the scale of this diagram.

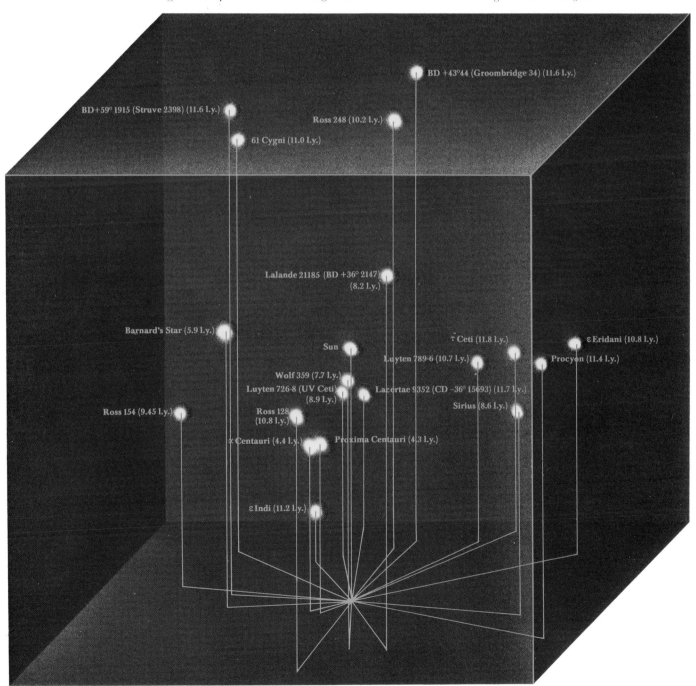

background scenery. This is obviously just a perspective effect, due to your seeing the finger from the different positions of the two eyes. Move the finger further away, and the shift between the views is less. You could in fact calculate the distance to your finger by measuring the angle it seems to shift through (and knowing the separation of your eyes). This is very similar to a surveyor's triangulation method, where he measures angles from each end of a baseline of known length to calculate the position of a distant object.

Turning to the stars, we can replace the finger by a nearby star, and the background by the scattering of very distant stars appearing near it on the celestial sphere. Photographs taken at different observatories should – we might think – show the nearby star at slightly different positions relative to the background stars. In practice, this shift is much too small to measure, because the observatories are too close together in comparison with the star's distance. A longer baseline is required, and the longest available to us is the diameter of the Earth's orbit. And by taking advantage of the Earth's movement in orbit – which establishes a base-line of 300 million km. (192 million miles) – astronomers can economize on observatories, using one instead of two.

A photograph is taken of the star concerned, and a second at the same observatory six months later, when the Earth is at the far point of its orbit. If the star is indeed quite close, careful

# Celestial Geometry to Gauge the Stars

Stars that are closer to us seem to move slightly against the background of more distant stars as the Earth swings round the Sun. The same effect can be created by holding up a thumb and opening first one eye, then the other. The point of observation is altered by a couple of inches and the thumb's position changes against the background. Both the change and the angle of the change are known as *parallax*.

The angle a star seems to move through as the Earth swings round the Sun *decreases* in exact proportion to its distance. Parallax angles (in astronomy, actually defined as *half* the apparent shift, i.e. the radius of the Earth's orbit) are measured in seconds of arc. One second (1″) is 1/60 of a minute or 1/3600 of a degree. A star with a parallax of 0.5″ would be twice as far as a star whose parallax is 1″. This system of measurement has given rise to the unit of distance known as the *parsec* (*parallax second*): one parsec is the distance at which a star has a parallax of 1″. In the example, this is the distance of the second star; the first would be at a distance of two parsecs.

One parsec is 31 million million kilometres (19 million million miles). A light-year, the distance light travels in one year, is 9.5 million million kilometres (5.9 million million miles). Both units of distance are used in astronomy, and they can be converted by the equation: 1 parsec = 3.26 light-years.

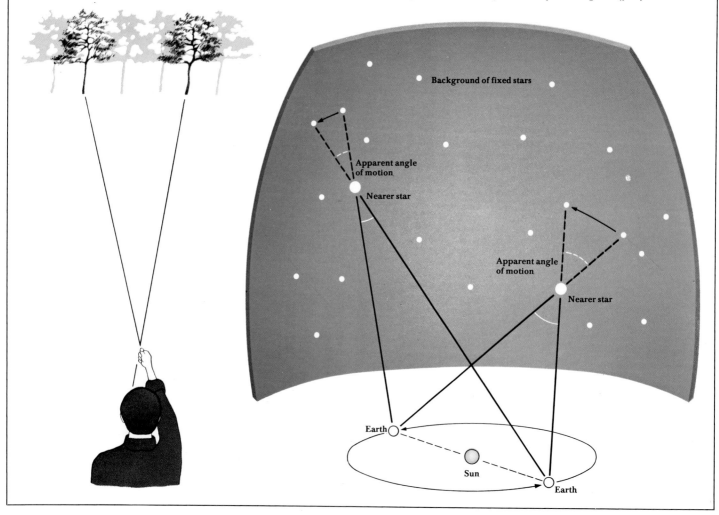

# Scales of Stellar Brightness

The brightness of heavenly objects is measured on a scale of magnitude. There are two kinds of magnitude:

—*apparent magnitude*, which is the brightness of stars as they appear to us on Earth; but since the more distant a star, the fainter it seems, apparent magnitude often bears little relationship to its —

—*absolute magnitude*, or its real brightness. Astronomers usually work in apparent magnitudes because that is the most useful guide to identification.

The Greek astronomer Hipparchus ranked the visible stars into six magnitudes – first magnitude for the brightest stars, and sixth for those only just visible to the eye. The stars, though, cover a whole range of brightness, and modern instruments can easily distinguish differences imperceptible to the eye. So Hipparchus's system has been tightened up, and magnitudes are quoted to the nearest hundredth part. Regulus, for example, has a magnitude of 1.36.

In this brightness scale, a difference of five magnitudes between stars corresponds to a brightness ratio of 100:1. The faintest stars visible to the unaided eye, at magnitude 6.5, are thus just over a hundred times fainter than Regulus.

Note that, following Hipparchus, the *brightest* stars have the *smallest* magnitudes. Sirius is so bright that its magnitude becomes a negative number. – 1.45.

At the other end of the scale, the telescope reveals stars fainter than magnitude 6.5, and these have correspondingly large numbers. Photographs taken with large telescopes can pick up stars as faint as magnitude 24 – about 10,000 million times fainter than Sirius.

---

measurements can reveal a slight shift in the star's position. Even so, the shift is small, and the measurement a correspondingly delicate one. To draw once again on the analogy of blinking at one's finger, the positional change of even the nearest stars is equivalent to the shift of a finger placed 10 kilometres from the eyes!

After decades of patient observations, parallaxes of thousands of stars are now on record. The distances of the nearest stars are accurately tied down, and the framework of the nearby stars is secure. Among the closest 20 stars we have Alpha and Proxima Centauri, our next-door neighbours in space; Sirius, a not particularly brilliant beacon in the celestial league, but the brightest star in our skies because it is only 8.6 light-years away; and Procyon, the bright star in the constellation Canis Minor, the Little Dog.

Most of our neighbours in space are dim creatures, though. Despite their proximity to us, they are visible only in a telescope. In real terms, then, they are shining extremely feebly, often with a thousandth or less of the Sun's light output. Since our region of space is undoubtedly quite typical, these faint stars must be the commonest kind in space. Although we think of the Sun as a typical star, it is in fact brighter than the average denizens of space.

When we push out our frontiers to the further stars, parallax measurements become less and less reliable. The further a star is, the smaller is its shift between the six-monthly photographs, and so the more difficult it is to measure its distance. If a star is more than about a hundred light-years away, its distance can be ascertained only within fairly wide limits. Yet a volume of space stretching out only a hundred light-years is only a fraction of stellar space. Fortunately, astronomers can supplement the parallax method by others, not so accurate for nearby stars, but with a greater range in space.

Most of these rely on the fact that stars are not stationary. All stars are moving, and with speeds which by ordinary standards are extraordinarily high – around 100,000 km. (60,000 miles) per hour which is several times faster than any spaceprobes launched from Earth. But because they lie at such huge distances, the stars' apparent movements across the sky – *proper motion* to astronomers – are very small. A nearby star like Sirius moves only 1/20th the Moon's apparent diameter in a human lifetime, and the further stars seem to be even slower. The celestial gold-medal sprinter, incidently, is the second-nearest star, a faint telescopic object called Barnard's Star, which moves through the Moon's apparent diameter ($\frac{1}{2}°$) in only 180 years. The constellation patterns are in fact gradually changing because of the stars' proper motions: the skies of 1,000,000 BC or AD 1,000,000 would be totally unfamiliar to us.

Although the stars' motions are small, modern techniques can pick them up readily and allow astronomers to follow the stars' tracks. Indeed, the yearly proper motion of a star is usually greater than its parallax, complicating the measurement of the latter. In practical parallax measurements, the star has to be followed during the course of the year for several successive years before the straight track due to its motion through space can be disentangled from the apparent 'wobble' in position caused by our changing viewpoint on the moving Earth. Without knowing a star's distance, its proper motion does not tell us how fast it is moving sideways through space – it could be a nearby, slow star, or a very distant one which has to travel phenomenally fast to cover the same angle in the sky in the same time. To measure its velocity we need to know its distance; or conversely, as we shall see, we can measure a star's distance from its proper motion if we have some independent way of knowing its actual speed across the sky.

The proper motion of a star also tells us nothing of how it is moving towards or away from us. By analysing the *spectrum* of a star's light, however, we can calculate precisely its speed along our line-of-sight, for the wavelengths of light are affected by the star's motion. By measuring this Doppler effect, the star's *radial velocity* can be worked out directly in kilometres per second. The star's real motion in space is evidently a combination of this radial velocity towards or away from us, and its velocity across the sky. A star travelling directly away from us will have no apparent velocity *across* the sky; one travelling at right angles to our line-of-sight (that is directly across the sky) will have zero radial velocity but a high proper motion. A star moving at 45° to the line of sight will have an equal amount of velocity directed along the line-of-sight (as radial velocity) as it has across the sky.

From these relationships between angle of motion, radial velocity, and proper motion, astronomers can measure star distances, albeit in a rather more roundabout way than the parallax method. But such techniques can be extended to stars whose parallax shift is too small to measure accurately.

Usually the actual direction of a star's motion through space is unknown, and astronomers must take an average over several stars which are intrinsically similar. It is however easy to measure the direction a *cluster* of stars is moving in. The stars in a cluster appear to converge towards a distant point, whose position depends on the cluster's direction of motion in space, just as the parallel lanes of a motorway appear to converge at the horizon. With a cluster of stars, then, astronomers can measure its distance without recourse to parallax at all. Distances to star clusters, measured by this *moving cluster method*, are very important in today's astronomy,

for they reach well beyond the range of parallax, and add all the members of a particular cluster to our list of stars with known distances. Some of these clusters also contain types of stars, like Cepheid variables, which are so rare that none are near enough for direct parallax distances.

From such a compilation of stars with distances accurately determined by one method or another, astronomers have been able to piece together a tremendously detailed picture of how stars work, how they are born and how they eventually die. The next chapter details this fascinating detective story of truly cosmic proportions; what is important here is that any two stars which are identical in particular details of their light (when examined spectroscopically) have exactly the same total output of light and other radiations (luminosity). Although stars cover a wide range in luminosity, from stupendous beacons a million times brighter than the Sun to glow-worms with a millionth the solar luminosity, as soon as

# The Message of Light

Ordinary light consists of a range of colours, as shown when it is split up as a rainbow. The spread of light according to colours is called a *spectrum*. When light passes through anything that bends the rays (whether a glass prism or a raindrop), the light separates according to the wavelengths it contains. 'Colour' is just our eyes' response to light of different wavelengths. Ordinary white light contains all the visible wavelengths, and the spectrum is a continuous bright band, stretching from violet to red.

In the early 19th century, the German physicist Joseph von Fraunhofer discovered that any chemical suspended in a gas absorbs its own particular wavelength of light. The absorbtion appears as a dark line – a sort of 'shadow' of the missing radiation – in the

spectrum. Fraunhofer mapped 576 such lines, and discovered that they were all associated with their own radiation frequency, and thus always showed up in the same position in a spectrum.

In a star, the outer layers contain many chemical elements and these absorb particular wavelengths of the star's light. When the light is spread into a spectrum, Fraunhofer's dark spectral lines – of which some 25,000 are now known – appear. Spectral lines are invaluable to astronomers. For a start, they 'fingerprint' each star and show what elements the star's atmosphere contains; in addition, a detailed study reveals its temperature, whether it is a giant or a dwarf star (and hence its luminosity), and whether it has a magnetic field.

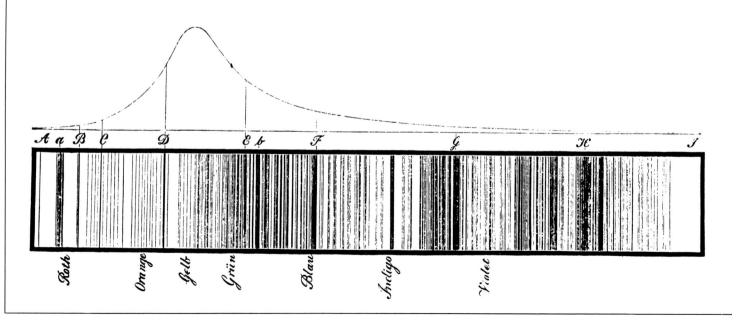

we have a star's spectrum we can match it with the spectrum of a nearby star whose characteristics are well known, and hence we deduce its luminosity. This star's apparent brightness in our sky depends both on its luminosity and how far away it is: the greater the distance, the more its light spreads out and weakens on the journey. The star's distance can thus be calculated by comparing its apparent brightness with its calculated luminosity.

So from our framework of local stars, whose distances are deduced from the parallax shift, and from nearby clusters of stars, with distances plumbed more indirectly, astronomers have interpreted the 'fingerprints' of star spectra, and from this scheme the distance to *any* star can be determined. All that is needed is its spectrum and a measurement of its apparent brightness. Such is the strength of modern astronomy that the third dimension of our star-filled night is firmly welded into place. Furthermore, the speeds of the stars as they travel their separate journeys is also open to the astronomers analysis. We are in a position now to describe the framework within which the stars exist and see how it fits into the entire scheme of the Universe.

**The Grand Design of the Stars**
Our eyes show us some 3,000 stars on a clear night – a total of 6,000 in the whole sky. Yet despite the impression of crowding we get on a truly brilliant star-lit night, these are only the nearest and brightest of a vast throng of stars. A telescope reveals multitudes of fainter stars, apparently increasing without limit. These distant stars are not scattered at random. They concentrate towards a broad band running right around the sky, and in the band itself so many faint stars are concentrated that, although each is individually invisible to the eye, the combined light makes a pearly glow as striking as the constellation patterns themselves. It, too, has been woven into

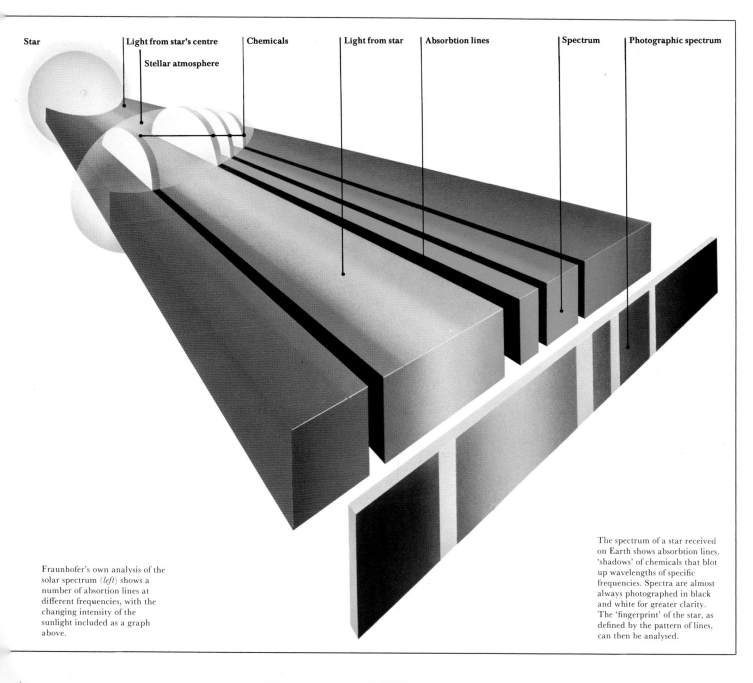

Star | Light from star's centre | Chemicals | Light from star | Absorbtion lines | Spectrum | Photographic spectrum

Stellar atmosphere

Fraunhofer's own analysis of the solar spectrum (*left*) shows a number of absortion lines at different frequencies, with the changing intensity of the sunlight included as a graph above.

The spectrum of a star received on Earth shows absorbtion lines. 'shadows' of chemicals that blot up wavelengths of specific frequencies. Spectra are almost always photographed in black and white for greater clarity. The 'fingerprint' of the star, as defined by the pattern of lines, can then be analysed.

legend. To the Greeks, it was milk split from the breast of Hera when she suckled the infant Herakles (Hercules); hence its modern name, the Milky Way.

The crowding of the stars towards the Milky Way is simply a perspective effect. The stars are not actually more closely packed there, but the stars of our system simply extend farther in that direction. The Milky Way's glowing band is telling us about the overall structure of the star system of which the Sun is one member, a star system named the Galaxy, from the Greek word for milk. Even before any star distances were known, 18th-century scientists such as Thomas Wright of Durham and the great astronomer Sir

William Herschel had realized the message of the Milky Way. Herschel counted the number of stars visible in his telescope in different directions in the sky, and deduced that if the stars were uniformly spread, they must be confined to a grindstone or lens-shaped system, with the Sun near the centre. When we look anywhere along the circle formed by the grindstone shape, we see light from many distant stars, which blend to produce the Milky Way; but looking up or down we see only the nearer stars with a dark sky behind, for there are no distant stars in these directions to give a background glow. Herschel's reasoning was entirely correct, but for other reasons his theory was dropped by 19th-century

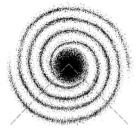

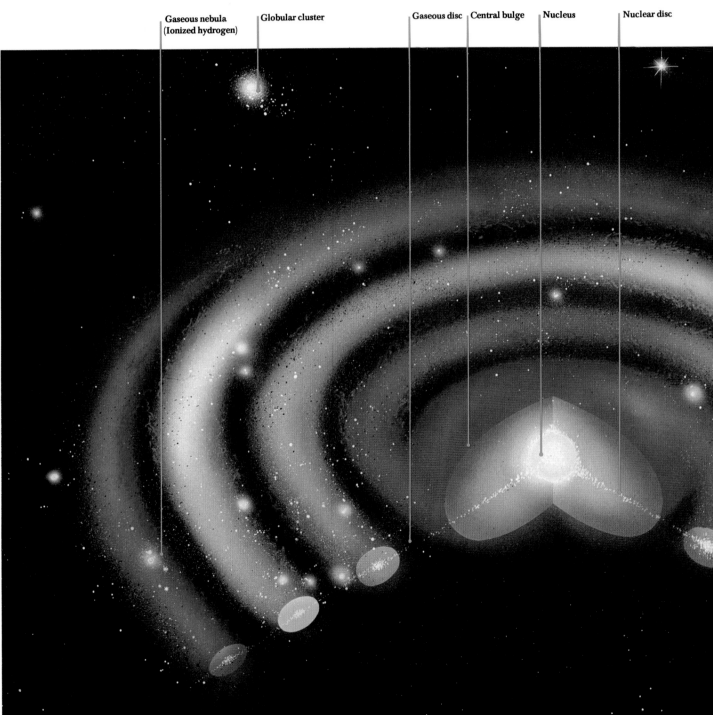

**Gaseous nebula (Ionized hydrogen)**   **Globular cluster**   **Gaseous disc**   **Central bulge**   **Nucleus**   **Nuclear disc**

astronomers. Its bones were resurrected early this century, and clothed in new observations to give us the true picture of our Milky Way Galaxy of stars.

It is indeed a lens-shaped star system. But the Sun is not central; it lies two-thirds of the way to one edge. Herschel and his successors were misled by a phenomenon unrecognized until the 20th century: the space between the stars is dusty. Small dust grains (mixed in with invisible gas) dim the light coming from distant stars, so that we cannot see even the centre of our Galaxy, let alone the 50 per cent of stars which lie on the far side.

The gravitational effect of all these stars,

whether or not our telescopes can detect them, is what shapes the Galaxy. In all, our Milky Way system consists of about 100 billion stars. Our Sun is just one member of a huge family of stars of all types, ranging from red giants hundreds of times larger to white dwarfs a hundred times smaller. Every star is pulling on every other by its gravity; and every star has its motion through space.

Yet all is not chaos. It we could stand back and look at the Milky Way system as a whole, we could see that it is a remarkably orderly system. Although our system is roughly lens-shaped, the modern view is not that it just tapers off in thickness as we go out to the edge. The outer parts of the Galaxy, where the Sun lies, actually constitute a thin disc whose thickness is roughly constant. Its proportions are similar to a gramophone record, but on an enormous scale: from edge to edge it measures 100,000 light years. This is a scale too tremendous to comprehend. Even the nearest star is four light years away, and if we imagine a scale model where Alpha Centauri is literally our next-door neighbour, the Milky Way Galaxy is a super-city entirely covering an area the size of Greece. And to be accurate, our city would also have to extend upwards several hundred metres to represent the thickness of the disc.

Towards its centre, the Milky Way system bulges up and down, like the yolk in the middle of a fried egg, its thickness increasing from 2,000 light years in the disc to about 10,000 light years at the centre. Before looking at the central regions, though, let us start by investigating the Sun's neighbourhood in the outer disc.

The framework of star distances, so painstakingly built-up, puts our region of the disc into perspective. At first sight it seems that the nearby stars are not going anywhere in particular: each is moving with its own random velocity, without any overall pattern. This is pure illusion. It is natural to think of the Sun as being at rest, but since our planet Earth will be carried along with it however the Sun is moving, this may not be so. And indeed indirect measurements show that the Sun, and most of the nearby stars are speeding through space at a rate of 250 km. (156 miles) per second.

All the stars in the disc of the Galaxy are revolving in orbits around the galactic centre, and the motion of the Sun and nearby stars is just part of this general merry-go-round. The orbits are kept in rein by the gravitational pull of the Galaxy as a whole – in other words, all the stars towards the centre contribute a gravitational tug which helps to hold the Sun in its path. Unlike the Solar System, whose planets are controlled by one massive central body (the Sun), the stars moving around the Galaxy feel the smoothed-out pull of billions of other stars. Despite the Sun's enormous speed, the Milky Way system is so huge that one complete orbit takes 250 million years to complete.

The random motions of the other nearby stars

Sun    Spiral arm    Globular cluster

A cut-away, three-quarter view of our Galaxy reveals its spiral structure, caused by the different speeds at which the various parts rotate. Our Sun takes 200–250 million years to complete one revolution. In the interstellar gas, which lies along the plane of the disc and between the arms, very hot stars create a scattering of hot-spots of ionized hydrogen. Around the outside of the Galaxy hang ball-like bodies of stars known as globular clusters.

can now be seen as a result merely of their slightly different orbits. A racing car driver sees the other cars slowly approach and recede from him as they speed on their slightly different paths around the track. Taking each car relative to the others, their common high speed does not matter: the deviations from parallel paths shows up as comparatively slow, random motions. And so it is with stars. Our neighbours are not pursuing exactly circular paths around the Galaxy's centre. They therefore have apparent motions which reveal themselves to astronomers as *proper motions* across the sky and *radial velocities* towards and away from us, but these speeds are much smaller than the stars' common velocity in orbit.

The Sun itself is not in a circular orbit. Its path deviates towards a point in the sky known as the 'solar apex', which lies near the bright star Vega. Sir William Herschel first measured this motion of the Sun, by averaging out the proper motions of nearby stars, but only in this century did astronomers realize that it is not a very important part of the Sun's motion. We know now it is just our stellar 'car's' sideways drift relative to the others on the course; far more significant is the fact that all the 'cars' are racing round the track at high speed in roughly the same direction. All the local stars, including the Sun, are heading in the direction of the constellation Cygnus.

The gravitational pull of the Galaxy's stars on one another gives the stars basically quite circular, orderly orbits. But there is one rather strange, and beautiful, result of this combination of gravitational attractions. The stars of the disc cannot stay uniformly spread out, but become clumped together in two spiral arms, which appear to wind out from the central nucleus. Deep within our Galaxy's disc as we are, it is difficult to see this pattern, just as the outline of a wood is not obvious when we are in amongst the trees. But there are millions of other galaxies, each – like our own Galaxy – consisting of billions of stars. Galaxies stretch out into the depths of space beyond the Milky Way, and in these distant galaxies we almost invariably find that the stars constituting the disc have clumped in a beautiful winding tracery of double spiral arms. Photographs overemphasize the appearance of the arms at the expense of the rest of the disc, and disc-galaxies like ours have been given the general name of *spiral galaxies*.

Although all the stars of our Galaxy's disc appear stacked behind one another in hopeless confusion, astronomers have been able to pick out local stretches of spiral arms relatively near to the Sun. The framework of star distances reaches out some 10,000 light years for the brighter stars, and spiral arms happen to be the abodes of the brightest stars in a galaxy. Astrono-

# The Galactic Halo

Knowing the distance from the Sun to the galactic centre is one of the most vital facts we want to know if we are to have an accurate model of the galaxy. But the distance cannot be measured directly. Although astronomers can assess the distances of many stars (by working out their absolute magnitudes and comparing this to their apparent magnitudes), they cannot see to the centre of our Galaxy, which is obscured from even the largest telescopes by interstellar dust.

The solution to the problem lies in the 'globular clusters', stars bound by their own gravity. These clusters – like M13 (right) – lie above and below the Galaxy's disc, in a halo, and they are not obscured by dust; and as they contain around a million stars each, even those on the far side on the Galaxy can be detected with a large telescope. We know by observing other galaxies that clusters are spread roughly uniformly around the galactic centre. By ascertaining the middle of the distribution of globular clusters, astronomers can find the position and distance of the galactic centre (above right).

American astronomer Harlow Shapley first fixed the centre of our Galaxy around 1920, using this method. The nearest globular clusters lie close enough to us for individual stars to be detected in them. Their distance can thus be measured directly. Shapley calculated the distances to the further ones by comparing their apparent

sizes with those of their nearby brothers, and worked out that the centre of the globular cluster distribution – and hence the Galaxy's centre – lies 50,000 light years away, in the direction of the constellation Sagittarius. Later research has refined Shapley's figure, reducing the distance to 30,000 light years. Our Sun lies about two-thirds of the way out from the centre towards the galactic edge.

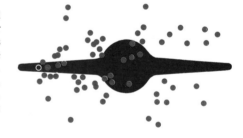

mers have located a spiral arm some 7,000 light years farther out than the Sun, running through the constellations Perseus and Cassiopeia, while towards the galactic centre another runs by at about the same distance, appearing in the constellations Sagittarius and Carina. These seem to be turns of the two major arms of our Galaxy. The bright, relatively nearby – on this scale – stars of the Orion region were once thought to be a local arm in which the Sun is placed, but astronomers now think it is merely a spur projecting from the inner edge of the ragged Perseus arm.

The disc – and especially the spiral arm regions within it – are where the action is in a galaxy. Stars are being born here, and stars are dying spectacular deaths as supernovae. These stories concern not just stars, though, but also the gas and dust lying between the stars, and the full story of the disc must take into account these insubstantial but vital components. Before leaving the basic framework determined by the stars, however, we should turn to the central parts of the Galaxy.

The nucleus is the geriatric ward of the Milky Way. No stars are being born here at the present time, nor have any since the original formation of the Galaxy. These stars are all around 15,000 million years old, three times the Sun's age, and they make up a fairly rounded bulge at the centre of the disc. The Galaxy's 'yolk' of old stars is some 10,000 light years thick and 20,000 light years in diameter, and its shape is a reminder that the Galaxy originally condensed from a colossal almost spherical gas cloud. The first stars to form from this gas did not travel in orderly circular orbits, but took elongated paths arranged higgledy-piggledy around the Galaxy's centre. Still pursuing these paths determined in their youth, the stars of the nucleus obstinately retain the original shape of the Galaxy, while the remaining gas smoothed out its turbulent motions and settled down as a smoothly rotating disc.

The original gas cloud was rather larger than the Milky Way system's present size. As it collapsed under its own gravity, the stars of the nucleus formed at the centre, where the gas was densest. But – for reasons which astronomers still do not really understand – condensations of gas further out collapsed individually to make *globular clusters*, each a closely packed ball of about a million stars. The 125 known globular clusters are spread out around our present Milky Way Galaxy in a huge spherical volume called the *halo*, the ghost of the original gas cloud. The two brightest globular clusters are actually visible to the unaided eye, appearing as 'fuzzy stars', and they were indeed originally catalogued in the same way as stars, by the names Omega Centauri and 47 Tucanae.

The halo region also contains a host of individual old faint stars. Their light output is so feeble that halo stars outside the giant globular clusters are difficult to detect. But these stars, like their contemporaries in the nucleus, are pursuing elongated, randomly orientated orbits around the galactic centre, and in doing so they must at some time pass through the disc. Some halo stars thus come within spitting distance of the Sun, and astronomers can study them if they can spot them amongst the multitude of disc stars. In fact, their orbits give them away. Since they are not partaking of the disc stars' circular dash around the Galaxy, these stars are being rapidly left behind. From our vantage point, they seem to be shooting backwards. Since these stars were identified before astronomers realized the Sun's high speed around the Galaxy, the slow-moving halo stars were dubbed 'high-velocity star' – and with the innate conservatism of astronomical names, this misnomer is still in use.

The old regions of the Galaxy – the nucleus and the halo – seem staid and unexciting. But in recent years, some astronomers have suggested that we have misjudged our Galaxy's halo. There are some indirect reasons for thinking it may contain a lot more mass than can be accounted for by the dim stars within it. If this suggestion – and it is only a suggestion – is correct, then the halo could contain more matter than the rest of the Galaxy. And this matter cannot be in the form of stars: it would have to be bound up in some dark type of object, perhaps a multitude of free planets, or even black holes.

The old region at the centre of the nucleus also has surprises in store. Radio astronomers have found evidence for an explosion, or series of explosions, there; and an intense central radio source. In some ways, the very centre of our Galaxy resembles – on a very much smaller scale – the explosions of the distant and enigmatic quasars. Later chapters will deal with the surprising violence that astronomers are finding in the hearts of many galaxies, so let us now return to our local neighbourhood, the Galaxy's disc, where star-life and death continues in a regular cycle.

**Between the Stars**

The yawning gulfs between the stars are by no means 'empty space': nowhere in nature is there a perfect vacuum, a region devoid of all matter. A scientist's 'vacuum' in the laboratory is not completely empty of gas, even though this residual gas is very tenuous in comparison with the crowding of molecules in air. Interstellar gas is a much better 'vacuum' than this, but it is still filled with extremely tenuous gas, mainly hydrogen, spread out so thinly that there is on average only one atom to every five cubic centimetres of space – half a dozen atoms in a small matchbox volume. Our Galaxy is so huge, though, that the total amount of gas between the stars in its disc would be enough to make 10 billion stars – in another words, 10 per cent of the matter of the Galaxy is not bound up in stars but is spread out as gas between them.

In the early history of our Galaxy, after the nucleus and halo stars had formed, the remaining

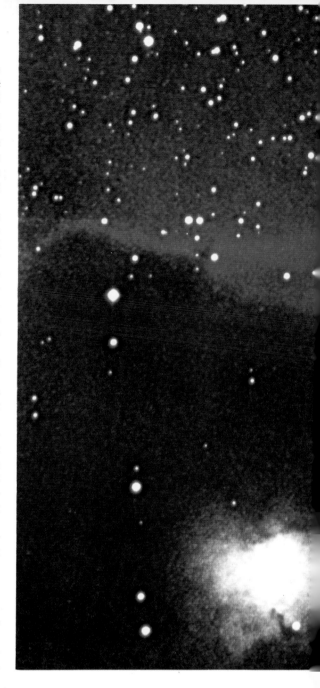

gas settled down into a rotating disc. Most of this gas formed into the disc stars, which still pursue the circular paths dictated by the earlier gas disc, but some of the gas has always remained. As we shall see in the next chapter, dying stars eject gas back into space; this mixes in with the original primordial gas, and eventually ends up in new stars. The disc is a dynamic place, with stars always being born out of gas, and in dying replenishing the surrounding gas reservoir.

The gas from a dying star, however, always contains slightly less of the hydrogen and helium which were the sole constituents of the original gas disc, and in their place contributes 'heavier' elements like oxygen, carbon and iron. The interstellar gas is thus gradually changing in composition, becoming 'enriched' in heavy elements. Some of these elements do not appear as gas, though, but as small solid grains of dust which obscure the light of distant stars.

To the traditional astronomer, observing the light from the stars, the dust is mainly a nuisance. Although there is a hundred times more gas than dust, weight for weight, the gas is transparent while dust is very efficient at blocking off light. The Milky Way in the region of the constellations Cygnus and Aquila, for example, seems split in two lengthways, but this is nothing to do with the distribution of the stars themselves. The culprit is the narrow layer of dust lying along the midplane of the disc, absorbing light from the distant stars near the central line of the Milky Way.

From southern latitudes, an even more prominent black cloud – the Coal Sack – stands out near the Southern Cross. Such dark clouds – and photographs reveal hundreds of them silhouetted against the Milky Way – show that the dust and gas in the disc are not uniformly spread. The distribution has been clumped into denser clouds, like the Coal Sack, by a variety of forces. The pressure of light from intensely bright young stars forces gas away from them, as does the explosion of an old star as a supernova, and in both cases a shell of denser gas piles up around the star. Changes in the surrounding gravitational field can also compress gas into a cloud, particularly when the interstellar gas is moving through a spiral arm of the Galaxy.

At the centre of a gas and dust cloud forced to contract in this way, the gas may become so compressed that its own gravity pulls it unremittingly together, and the gas condenses into a cluster of stars. Radiation from these young stars will eventually blow the residual gas away, but first it makes the gas glow. Ultraviolet radiation is mainly responsible. Just as teeth, or a freshly laundered shirt will shine in the 'black light' ultraviolet lighting of a party or discotheque, so the gas in the clouds glows in the radiation from the central stars. The most famous example of such a bright *nebula* (a word meaning cloud) is the great Orion Nebula, 15 light years across, and lit up by four young stars in a central group called the Trapezium.

Nebulae and the bright stars lighting them occur throughout the Galaxy's disc, although they are most common in the spiral arms. They are some of the most beautiful objects in the skies. A small telescope shows the wisps in the Orion Nebula, while long exposure photographs reveal hundreds of beautiful nebulae, in a fascinating variety of shapes. Their outlines are determined not only by the edge of the gas cloud itself, but by the extent to which the radiation from the central stars can reach. If dark clouds on the near side are out of the radiation, they will block light from the nebula behind, producing fantastic black silhouettes, often in the form of long conical 'elephant trunks'. The most famous dark patch in a nebula is the Horsehead Nebula in Orion (not far from the Orion Nebula), an uncannily realistic equine head silhouetted against a diffuse background glow. Dust patterns stretching over whole nebulae have produced glowing clouds in such forms as the trisected Trifid Nebula, the cartographically perfect North America Nebula, and the swirling Lagoon Nebula.

The message of light from space tells us not very much about the transparent interstellar gas itself, nor about how stars form deep within the obscuring cocoons of a nebula's dark dust clouds. Today's astronomers are however not limited to investigating merely the light from space. In Chapter 9 we shall find how other radiations from space, ranging from radio and infrared to ultraviolet, X-rays and gamma rays can be picked up and analysed. Since modern physics can tell us what is emitting the radiations, astronomers have effectively opened several more eyes on the Universe.

For the investigation of interstellar space, radio waves and infrared have the enormous advantage that they can travel unimpeded through the dust. Thus we can pick up radio waves from natural transmitters right the other side of the Galaxy. Natural radio broadcasters take several forms, but the most interesting for astronomers probing the Galaxy is hydrogen gas. A hydrogen atom broadcasts quite spontaneously

at an exact wavelength of 21.1 cm. Since hydrogen is the commonest gas in interstellar space, a radio telescope tuned to 21 cm. receives a cacophony from hydrogen atoms throughout the disc of the Galaxy.

Astronomers can however interpret the message of the 21 cm. radiation. Like the spectral lines in a star's spectrum, the 21 cm. radiation is slightly displaced in wavelength if its source is moving towards or away from us, because of the Doppler effect. Radio astronomers use this effect to measure how fast the gas is moving throughout the Galaxy's disc. The gas is orbiting under the gravitational influence of the stars, and by investigating the orbital velocity of the hydrogen in different parts of the Galaxy, astronomers can probe the Galaxy's gravitational field at distant points.

Once we know the orbital speed of the disc at different distances from the galactic centre, we can turn the argument round and determine the distances of individual hydrogen clouds from

The Horsehead Nebula in Orion is a dark cloud that obscures lighter regions of hotter gas and countless stars. The stars in the lower half of the picture are in front of the cloud. At lower left in the panorama is another, hotter gas cloud reflecting the light of a hidden star.

The Pleiades 'open cluster' – a young group some 60 million years old – is a dense mass of stars still in the process of formation. The gas that surrounds them like a fog will eventually be dispersed. The group is commonly known as the Seven Sisters, because that is the number of stars usually visible to the naked eye. But long-exposure photography has revealed that there are several hundred stars in the group, which will gradually break up as its members age.

their measured radial velocities. In this way, radio astronomers can provide distances to radio sources anywhere in the Galaxy, well beyond the optical astronomers framework.

Moreover, interstellar gas is more concentrated in spiral arms, so the arms emit stronger radio signals. By measuring the distances to the more powerful hydrogen clouds, radio astronomers have mapped the entire spiral structure of our own Galaxy, showing it to be very similar to other spiral galaxies.

These radio observations have also shown that the interstellar gas between the obvious nebulae is certainly not uniform – unlike the gas of the Earth's atmosphere which is all at roughly the same density and temperature (at sea level). Travelling through interstellar space, a voyager of the future could find himself passing through regions a hundred times denser than average, and at a temperature of $-200°C$, and in a few light years emerge into a stretch of space filled with gas a hundred times thinner, at $3,000°C$, (though the very low density of gas everywhere in space means that his spacecraft would not actually be frozen and heated).

Threading through all the gas in the disc is an enormously extended magnetic field, 100,000 times weaker than the Earth's field. Astronomers have mapped the Galaxy's magnetism by its effect on radio waves from distant sources; and by its effect on the dust grains in space.

The final components of the disc are the insubstantial 'cosmic ray' particles. These are electrically charged fragments of atoms speeding along at almost the velocity of light. Nine in ten are protons, the nuclei of hydrogen atoms, and the rest mainly helium nuclei, while one in a hundred is an oppositely charged electron. Astronomers believe that these atoms have been shattered and the fragments flung up to enormous speeds by the colossal explosions of supernovae, either in the explosion itself or else by the ultra-energetic pulsars left afterwards.

At their enormous speeds, cosmic rays should quickly shoot out of the Galaxy, but the magnetic field bottles them in, so that they keep retracing their paths in the disc. As a result, some penetrate the Solar System and can be studied from the Earth or from space probes travelling between the planets. Astronomers can also investigate cosmic rays far out in the disc, for as these electrically charged particles gyrate around the lines of magnetic force they generate radio waves. Unlike the 21 cm. broadcast of hydrogen, these *synchrotron* radio waves come out at all wavelengths, like noise all the way along the receiver

single atoms. Only sheltered within an ultra-violet-absorbing umbrella of dust can the molecules survive intact.) Unfortunately, hydrogen molecules are not radio broadcasters. Astronomers were long in ignorance as to exactly how much hydrogen there is in the centres of clouds, let alone its temperature and density. The answer came in the 1960s by an indirect route, when radio astronomers picked up radiations of precisely defined wavelengths, which were *not* 21 cm. These radio waves are emitted by a variety of molecules. First identified was hydroxyl, a bound oxygen-hydrogen pair, which is essentially water with one hydrogen atom missing. Other discoveries soon followed, and the total number of molecules found in space now stands at around fifty. The largest molecule presently known contains a string of 11 atoms, and familiar compounds so far identified include carbon monoxide, formaldehyde (the medical preservative) and ethyl alcohol. Although molecules are very rare in space when compared to hydrogen, itself very tenuous by Earth standards, the size of astronomical objects means that the total amount of these compounds is enormous in everyday terms: an interstellar dark cloud contains enough alcohol to fill the Earth, were it hollow!

The vast majority of these molecules are found only in dark clouds, where the dust protects them from being broken down into atoms by the ultraviolet radiation in space. They probably form on the surface of the dust grains, where atoms can come together and combine into molecules.

In the cloud centres, the gas is packed a million times more tightly than it is in the general interstellar medium, and these are just the right conditions for stars to form. As we shall see in the next chapter, even the first stages of a star's life produce a great deal of light and heat, which is trapped by the surrounding dust, producing infrared ('heat') radiation which can penetrate the dust and escape from the cloud.

Later on, these stars will disperse the dust around them, and light up the surrounding gas as a nebula. When the nebula disperses, the stars will be left as a cluster, like the well-known Pleiades (Seven Sisters) cluster in Taurus. The Galaxy's disc contains an estimated 10,000 such 'open' clusters, all formed relatively recently on the cosmic timescale. The Pleiades, for example, are 'only' 60 million years old, born at the time the dinosaurs perished on Earth.

Unlike the old compact globular clusters of the halo, the disc's open clusters are only loosely bound together by their own gravitation, and will eventually disperse as individual stars. The Sun, and its family of planets, must have begun in such a cluster some 4,600 million years ago, though the Sun's siblings have long since dispersed to other parts of the disc. The cluster stars spreading out to populate the disc will eventually die and return some of their gases to space to continue the cycle of star life and death, which we shall follow in the next chapter.

band of a home radio in contrast to a broadcast at a specific wavelength. The newest tool in the study of cosmic rays are the gamma-ray satellites. When the superspeed cosmic rays smash into stationary hydrogen atoms in space, they emit the very short wavelength gamma-rays. By mapping the directions from which gamma-rays originate, astronomers are learning about the distribution of cosmic rays in the farther reaches of the Galaxy's disc.

And all this does not take into account the dense gas clouds, found especially in the spiral arms. Deep within these clouds, hidden from the optical astronomers by wreaths of dust, stars are forming. Modern astronomy is now laying bare the process of star birth itself – invisible in ordinary light – and revealing the stages between the collapse of gas as a dark cloud and the spectacular unveiling of the young, fully formed stars at the centre of a nebula.

The clouds of particular interest to astronomers, for in them are formed surprisingly complex molecules. In the centres of the clouds, hydrogen atoms combine in pairs to make hydrogen molecules. (These hydrogen siamese-twins are the usual form of hydrogen we encounter on Earth, but in ordinary interstellar space the energetic ultraviolet radiation splits them into

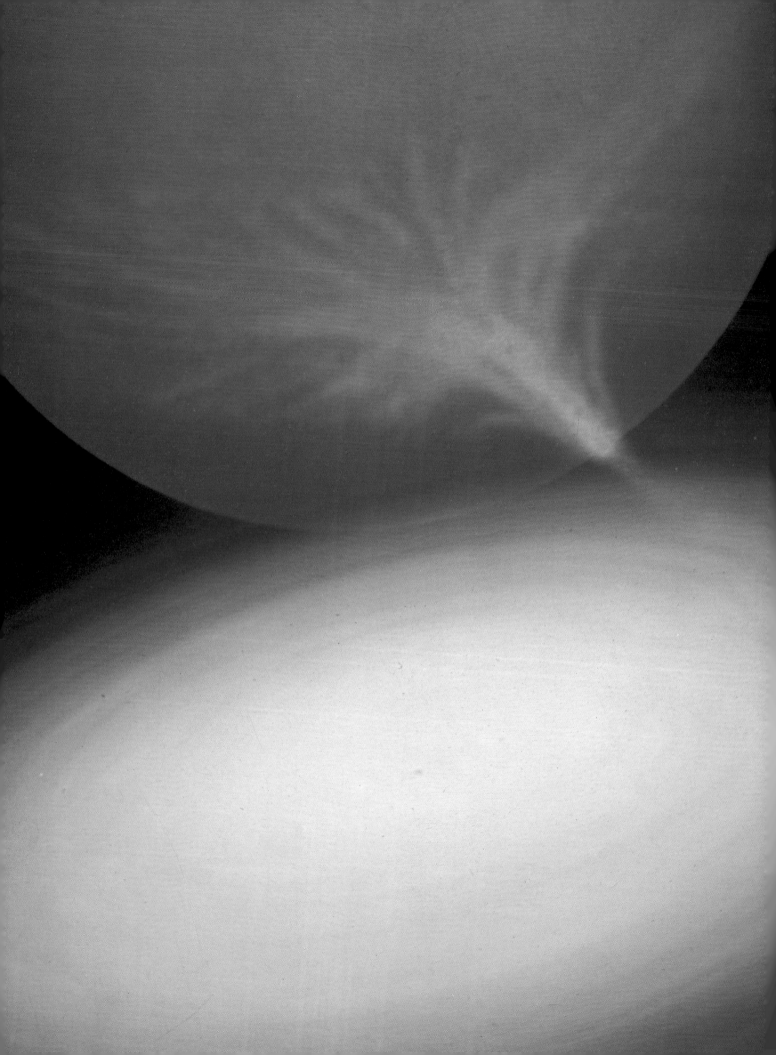

# 5/STELLAR EVOLUTION

Stars are the crucibles of the Universe. Each tiny point of light in the night sky is a furnace, fuelled by the same sort of gas that tenuously pervades the voids of interstellar space. Burning with the energy of countless hydrogen bombs, it is the stars that from their fuel form almost all the heavier elements familiar on Earth. Indeed, the material that forms us and all you can see around you was once inside a star, or perhaps many stars. These elemental furnaces have many different forms. Some are small, burning quietly and unnoticed. Some end their lives in violent explosions, blasting their gases back into space. Some are so massive that they collapse under the force of gravity into super-dense black holes, swallowing matter swept from any neighbouring star (left) and keeping captive even their own light.

**S**tars are the basic cells of the astronomical world. They are the furnaces in which the future of the Galaxy and other galaxies is being forged, for stars create almost all those elements that are more complex than hydrogen and helium. This building up of heavier elements is slowly changing the chemical composition of the Universe as gas is processed through stars and returned to the interstellar medium. Not only is the chemical state of matter changed in this way, but the very process of creating heavier elements provides the energy which gives life to stars and makes them luminous objects.

Stars originate in interstellar space itself. To the naked eye, the space between the stars appears to be an empty void. In one sense this is true. Most of interstellar space is a vacuum more rarefied than any that can be produced in a laboratory on earth. But space is not totally empty. It contains atoms, dust and electro-magnetic radiation.

The typical distance between individual atoms in interstellar space is about a centimetre: on average there is only one atom in a cube the size of a sugar lump. However the size of the Galaxy is so vast that even this extremely low density gas contains an enormous mass of matter.

It is important to realize how enormous this mass is. The gas lies in the plane of the galactic disc in a thin sheet about 300 light years thick, less than one two hundredth of the radius of the Galaxy. The volume of the gas is equivalent to $10^{67}$ sugar lumps, that is $10^{67}$ cubic centimetres, each cubic centimetre containing one atom on average. The total number of atoms in the gas is therefore $10^{67}$. The gas is composed mainly of hydrogen, each atom of which has a mass of about $10^{-24}$ g. (one million-million-million-millionth of a gram). Hence the total mass of gas in the Galaxy is about $10^{-24} \times 10^{67}$, that is $10^{43}$ g., or about 10,000 million ($10^{10}$) times the mass of our own Sun! Since the Galaxy contains about $10^{11}$ stars, each typically with a mass something like that of the Sun's, we see that about 10 per cent of the material in the Galaxy is in the form of gas – hydrogen and helium, with a small admixture (about two per cent by mass) of heavier elements (principally carbon, nitrogen and oxygen).

In addition to this gas, the interstellar medium contains minute grains of solid dust particles. These dust grains are extremely small, less than one hundred-thousandth of a centimetre across. The presence of this fine dust can be deduced from the effect it has on the light of stars shining through it. The dusty regions of space have two effects on observations. Firstly they cause the stars behind them to appear fainter than they would if that light travelled unhindered through the Galaxy. The reason for this is that starlight is scattered and absorbed by the dust and is thus reduced in intensity. Hence stars at larger and larger distances are not only fainter because of their distances, but also fainter because they shine through increasingly thick layers of dust.

The scattering of starlight has a second, subsidiary effect. Just as the sun appears redder when seen through the smoke of an evening bonfire, so the light of stars is reddened by passage through the dusty regions of space. The reason for this is that, although all light is scattered when it shines through clouds of dusty material, the red light from a star is scattered less than the blue light, and so the observer sees more of the red light and the star appears redder. Conversely, if a particular dense cloud of dust lies near a star, the blue light which it preferentially scatters will cause the cloud to glow as a faint blue nebula.

The interstellar medium between the stars also contains very high energy particles – rapidly moving electrons, protons and atomic nuclei – called cosmic rays. These cosmic rays are charged particles, and spiral round the weak magnetic field lines in the Galaxy. This spiralling motion generates radiation at radio wavelengths thus producing a diffuse source of radio emission throughout the Galaxy, with strongest emission in the plane of the Milky Way.

## The Birth of Stars

The gas and dust between the stars is not spread uniformly throughout the galactic disc. Instead it is clumped in a largely chaotic way, with concentrations along the spiral arms in the form of clouds and nebulae. It is this deviation from a uniform distribution that triggers the birth of stars. For, once a deviation from a uniform spread of gas is large enough, the force of gravity takes command. The subsequent evolution depends on the balance between gravity and internal pressure generated by heat: the hotter the cloud, the greater the outward push of pressure; the more massive the cloud, the greater the inward pull of gravity. Small, low mass and hot clouds tend to expand like a balloon being blown up, with pressure dominating gravity, whereas large, dense and cool clouds collapse like a balloon deflating.

At the temperatures and densities typical of the gas clouds in the Galaxy the critical mass which is required for collapse to occur, known as the Jeans mass, after the British physicist Sir James Jeans, is several thousand solar masses. If collapse were to occur without further changes in the cloud, stars would be formed which were several thousand times more massive than the Sun. However, once started, the process of collapse, which takes many millions of years, becomes unstable. The cloud breaks up into sub-condensations. The increasing density of the initial cloud as it contracts leads to an increasing gravitational force. Since it is transparent in these early stages it can leak energy out and thus remain cool. The outward push of pressure remains the same but the gravitational pull gets stronger. The critical Jeans mass decreases and the cloud fragments into smaller and smaller sub-clouds.

This break-up into smaller fragments cannot continue indefinitely. Eventually the density of

The dense mass of stars crammed into the disc of our Galaxy testifies to the immense bulk of matter from which such stars condense – and to which they will finally return most of their gases.

each sub-cloud rises to a point where dust grains start to trap heat. The temperature, and therefore the pressure, rises, and pressure starts to win the battle against further fragmentation. The sub-cloud has then reached its final stable mass and becomes a protostar.

From this point onwards each protostar evolves as a separate coherent object. The inner regions contract to form a hot nucleus and the outer regions settle onto this central core over the next million years. By this stage each protostar has become opaque, and heat can no longer leak out easily. As the protostar continues to contract so its temperature increases steadily. Starting as a cool distended object with a surface temperature of 1,000° radiating strongly in the infra-red region of the electromagnetic spectrum, it contracts and increases in temperature to become gradually brighter at visible wavelengths.

What stops protostars from contracting indefinitely? The barrier which eventually halts this contraction is thermonuclear fusion. The same source of energy that is liberated in the hydrogen bomb acts to stabilize and halt the collapse of protostars. At a temperature of about 10 million degrees centigrade in the centre of the star, nuclear fusion of hydrogen into helium starts to occur. The energy released in these fusion reactions keeps the central regions hot, and continually replenishes the energy which is slowly leaking out through the star. In this way a balance is achieved between gravity pulling inward and the pressure pushing out, a balance which can continue to be maintained so long as thermonuclear fuel is available. The stage when nuclear 'burning' starts to liberate thermonuclear energy is the stage when a star is truly born – a cool, transparent cloud has fragmented into a cluster of a hundred shining stars.

## Stars in Middle Age

The newly-formed clusters of young stars are visible in the Galaxy as associations and open clusters. Because they are born with random motion, the stars in a cluster will slowly disperse,

spreading themselves throughout the Galaxy. As we shall see, more massive stars, with masses 20 or 30 times that of the Sun, cannot survive longer than a few million years before consuming the bulk of their nuclear fuel. In this time, with the stars moving off on their separate trajectories, the typical cluster will be broken up. Thus the massive O and B spectral class stars tend to be found mostly in open clusters. By the time the cluster has dispersed these stars have already reached the end of their life.

Clusters containing young, massive O and B stars are found more frequently in the spiral arms of the Galaxy. For this reason it is believed that the spiral arms act as triggers for star formation. Although the Galaxy looks as if it should be revolving like a giant Catherine-wheel, its motion is not as simple as the image suggests. The gas, dust and stars in the spiral arms do not move round the Galaxy at the same rate as the spiral arms themselves. Rather, the arms are thought to be waves, moving around the Galaxy rather like water waves moving across the sea. But in contrast to water waves, the spiral arms move around the Galaxy more slowly than the gas and stars. The passage of these arms disturbs the interstellar medium and generates the inhomogeneities – the regions of denser gas – that induce the collapse of interstellar clouds. In this way young stellar clusters are formed largely in the arms of the galactic disc.

All the stars in one of these young clusters are born at the same time, from a gas cloud of uniform chemical composition. The only major difference between them is their mass. A whole range of stellar masses are formed, varying from the lowest masses of less than a tenth of the mass of the Sun, up to the highest of about 50 solar masses. However, the process of fragmentation leads to many more low mass stars than high mass stars being formed.

The intrinsic brightness of these stars in a single galactic cluster also varies from one to another, but in a way which is found to be closely correlated with their mass – the more massive stars are brighter. Furthermore the brightest stars are also hottest, with a surface temperature of 40,000°C. They are bluish O-type stars. The fainter stars are quite cool, going down to temperatures of 3,000°C. Their spectral appearance is again different: they are faint 'red dwarfs'.

The brightness, or luminosity, of a star is normally measured in units known as 'absolute magnitudes' (i.e. the star's actual brightness, rather than its 'apparent magnitude', which varies according to the star's distance from us). If the absolute magnitudes of all the stars in a young open cluster are plotted against their

surface temperatures, the relation between brightness and temperature is shown up in a striking fashion. All the stars lie in a narrow band extending from the hot, bright stars in the top left-hand corner down to the cool, faint stars in the bottom right-hand corner. Stars in this band are said to lie on the 'main sequence', and such a diagram is called a Hertzsprung-Russell diagram after the two astronomers who independently emphasized its importance as a technique for studying stellar properties. The Hertzsprung-Russell diagram plays a major role in stellar evolution theory as a code containing all the main clues to the way stars evolve.

The Hertzsprung-Russell diagram relates the inherent brightness, or absolute luminosity of stars to their colour. Bright stars are shown at the top, dim ones at the bottom. Hot blue stars are on the left, cool red ones on the right. (The main spectral types are classified by letters often recalled by the mnemonic: O, Be A Fine Girl; Kiss Me.) Most stars show a close link between colour and luminosity, and fall on a line known as the 'main sequence'. Other types are abnormal. Some, although cool, are very bright and red as a result of their giant size. Others are dull and small, but with a high surface temperature. These are ancient, dying white dwarfs. The line shows the probable evolution of the Sun, first to a red giant 10–100 times its present size and finally to a white dwarf one-hundredth its present size, all its energy spent.

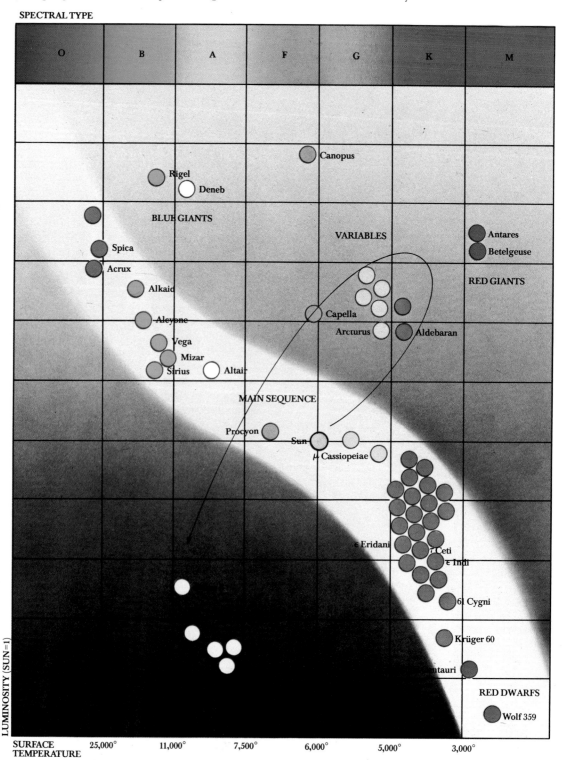

Like the stars in an open cluster, and indeed like the majority of stars in our Galaxy, the Sun is a main-sequence star, and sits right in the middle of the main-sequence band. It has an average surface temperature, and an average brightness. But the Sun has not always been the way we see it now. Five thousand million years ago the Sun was born as a result of some slight disturbance of the interstellar medium – a neighbouring supernova explosion or the passage of a spiral arm perhaps acting as the trigger.

The main sequence stars have reached the adult stage of the stellar life cycle. Once nuclear burning reactions commence, a star settles down into a stable, quiescent state, steadily converting its hydrogen into helium. Its brightness and surface temperature depend mainly on its mass, and will remain unchanged for a very long time. (For example, our Sun has been burning much as we now see it for some 5,000 million years, a little longer than the age of the Earth; it will go on burning for about another 5,500 million years without much outward change.) There the star will sit during this phase of its life, in the same position on the main sequence band of the Hertzsprung-Russell diagram. Only when the hydrogen fuel begins to be burnt out in the centre will any fundamental change take place.

In order for a star to remain on the main sequence three conditions must be satisfied. Firstly, there must be an energy source. Secondly, this source must supply energy at just the right rate to balance the energy which is leaking out from the hot interior, and which eventually escapes as the radiation we observe at the stellar surface. And thirdly, the pressure force pushing outwards must balance the gravitational force pulling inwards. The internal structure of a star is determined by the need to satisfy these three requirements.

The nuclear reactions that provide the energy work in the following way. Hydrogen nuclei fuse to form a helium nucleus, liberating energy in the form of high energy photons (or gamma rays) and neutrinos.

The neutrinos are massless particles and are basically unaffected by gravity. Furthermore neutrinos are most elusive particles, for they interact only very weakly with the other elementary particles. Once produced in the centres of stars they escape freely, disappearing into space at the speed of light.

Photons, on the other hand, though also massless and travelling at the speed of light, interact strongly with the atoms and electrons of the gas in which they are formed. Unlike the atoms and electrons, the massless photons are virtually unaffected by gravity. They are not bound to the star in the same way as the gaseous material, but, like neutrinos, try to fly out into space. However the outward flow of photons is hindered by the gas itself. A photon streaming out from the centre trips up in its flight whenever it hits an atom, which may then absorb it. One of the electrons encircling the atomic nucleus

then bursts out, flying off with some of the photon's energy. The higher the energy of the photon, the faster the atom and electron fly apart. In this way the energetic photon emitted by the nuclear furnace gives up its energy to the gas, keeping the centre of the star hot.

However, the electron itself will soon get captured, jumping back into an orbit around some other atomic nucleus and in turn releasing a new photon. Thus the gas of atoms and electrons engage in a complicated dance, sometimes absorbing photons as they fly apart, sometimes emitting them as they jump together. It is not the same photon which is involved in each collision and re-emission by an atom. Each photon is like a runner in a relay race, passing energy out like a baton after every interaction with an atom. The gas itself is held firmly in place by gravity, the atoms and electrons dashing about in such a way that there are always the same number on average at any one point in the star. However the photons do slowly drift outwards to the surface. Here they can escape freely as the light which causes a star to 'shine'. In this way the energy being lost from the surface of a star in the form of the radiation which we see, is being continually replenished by the nuclear furnace in the centre. What we see as points of light in the depths of the night sky are photons that have bumped and bounced their way out through the gaseous matter of the stars.

These interactions slow down the progress of the photons. If the photons emitted in the burning region of the sun were to escape freely they would travel out at the velocity of light, escaping from the surface after a couple of seconds. Instead it takes this radiation something like 50 million years to batter its way out. The energy we see escaping from the Sun today was first released when four hydrogen nuclei fused to become a helium nucleus some 50 million years ago, at a time when Earth's primates were emerging.

We have seen one way in which energy leaks out through a star – by the blundering walk of photons. Energy can also be carried out by another mechanism: by the motion of the gas itself. The hotter material at lower depths in a star may suffer from what is called 'convective instability'. Hot blobs of gas may find themselves surrounded by cooler, denser material. Because the hot blobs are lighter, they will rise up like bubbles of air rising in water. Only when they reach a place where their density is the same as that of their surroundings do they slow down, come to rest and release their heat. Similarly, cool blobs will be travelling downwards, in such numbers that on average the star neither expands nor contracts. In this turbulent region heat is physically carried by the gas from the inside to the outside of the convectively unstable region.

Stars differ in their structure depending on their mass. Theoretical studies show that the central nuclear burning region of a massive, main-sequence star is extremely compact, producing a small, high temperature centre. The

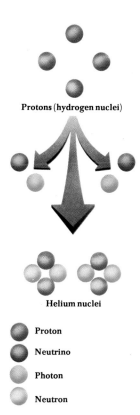

Protons (hydrogen nuclei)

Helium nuclei

● Proton
● Neutrino
○ Photon
○ Neutron

In the nuclear furnaces at the heart of stars, the nuclei of hydrogen atoms (protons) are fused in a series of complex reactions (here much simplified) to produce helium nuclei. In the reactions, energy in the form of neutrinos and light (photons) is released.

temperature drops rapidly away from the centre. The rapid drop in temperature makes the centres of massive stars convectively unstable. But their turbulent cores are surrounded by stable layers of gas. Photons slowly leak out through this quiet outer region to the surface.

Lower mass main-sequence stars, like the Sun, are different in every way. Lower mass stars have less extreme temperature changes outside their burning regions and are quite stable in their centres. But the outside is in violent motion. Low mass main-sequence stars are relatively cool, with surface temperatures less than 10,000°C. Below this temperature radiation finds it hard to escape because the stellar gas is opaque, rather like a very thick fog. With the photons dammed up in this fog, convection takes over as the most effective means of transporting energy outwards. So the outside of these stars is in rapid, turbulent, convective motion, with plumes of gas rising and falling as they carry heat out from the interior.

Our conclusions about the outside of low mass main-sequence stars are confirmed when the Sun is examined by telescope. The Sun is the only star whose surface we can examine closely. All the other stars remain as unresolved points of light through even the largest optical telescopes. The Sun, on the other hand, is so close that we can see details like sunspots even with the naked eye (though the sun *should never be looked at directly* as this seriously damages the retina of the eye). The Sun is seen to be covered by a fine cellular pattern of light and dark regions, of bright spots surrounded by darker dividing walls. The whole mass of cells, referred to by astronomers as the solar granulation, is seething, like a bubbling basin of rice pudding. These cells are the topmost layers of the convective zone that envelops the inner regions of the Sun.

What about the insides of stars? Is there any way in which we could directly confirm our conclusions about what is going on now in the nuclear furnace of the Sun? Since it takes 50 million years for radiation to escape, and the photons which escape are totally different from the photons which were first released in the central thermonuclear reaction, this would appear impossible. But there is another source of information: those elusive neutrinos. The neutrinos fly out unaffected by the surrounding gas and

Inside massive stars, nuclear reactions in the core create an unstable region of gas – with temperatures of many millions of degrees centigrade – which circulate the heat. This is then radiated out through a stable layer of gas. High mass stars burn their hydrogen in a mere 10 million years.

Convective gases

Core

Corona

Radiation

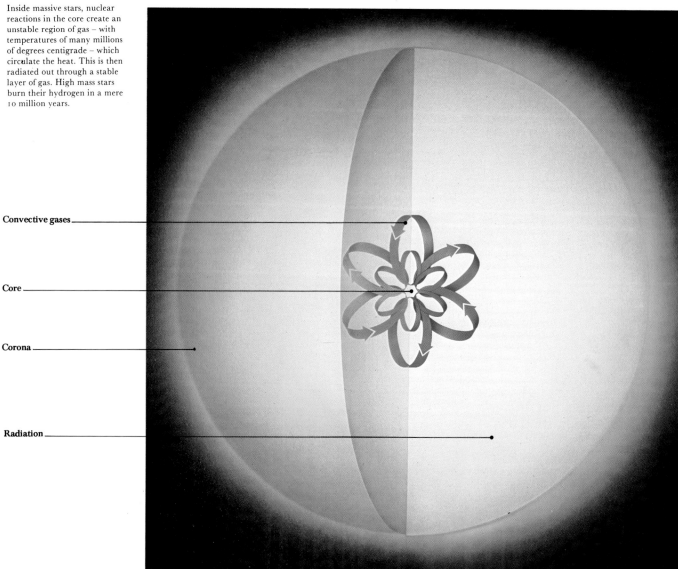

disappear into space. If they could be detected, the neutrinos would tell their story about the nuclear burning fires where they were created.

Detailed analysis of the nuclear fusion reactions occurring in the Sun predict a vast flood of neutrinos. Something like 100 million million are passing through this page every second. An American astronomer, R. Davis Jnr. has been searching for these neutrinos in a most extra-ordinary place – at the bottom of the Homestake gold mine in South Dakota, USA. The neutrino 'telescope' is a huge tank, containing several hundred thousand litres of dry cleaning fluid, perchloroethylene. This is a compound of carbon and chlorine; some of the chlorine is in the form of the isotope (an atom of different structure but the same chemical properties) chlorine 37. This isotope reacts weakly with neutrinos and is converted into the gas argon. The tank of fluid is kept a mile below ground to shield it from bombardment by other sources such as cosmic rays, which produce unwanted signals in the detector. By counting the tiny number of argon atoms produced in the tank, Davis is able to determine whether the Sun is emitting the number of neutrinos that theory predicts.

It is not! At the moment the sun is producing something like one-third of the neutrinos it should be. This is very worrying for theoretical astronomers, who thought they understood the structure of the Sun. Some of the details of the nuclear reaction calculations used to construct theoretical models of the Sun, and of stars in general, seem to be wrong, or else more complex processes are occurring than we presently understand. Though the general basis of stellar structure is well understood, and the properties we have described so far are well established, the solar neutrino experiment is a disturbing reminder that nothing in science can be taken for granted.

## Old Age

All stars are mortal. Indeed nothing in the Universe is unchanging except absolutely cold, dead, stellar remnants. Stars sitting on the main sequence are doomed to die because their reserves of energy cannot last forever. As they eat up the hydrogen in their centres they slowly exhaust the fuel which allows them to shine.

How fast a star runs through its adult life on

A mile down in the Homestake Goldmine, South Dakota, shielded from all normal cosmic radiation, this huge tank, containing 400,000 litres of dry-cleaning fluid is used to trap elusive neutrinos, produced in the interior of the Sun.

A low-mass star, like our Sun, burns its hydrogen over 10,000 million years. The central regions, at relatively low temperatures, are stable, and radiation leaks away from the core until blocked by the outer gases, which circulate like boiling oil to release the star's heat at the surface.

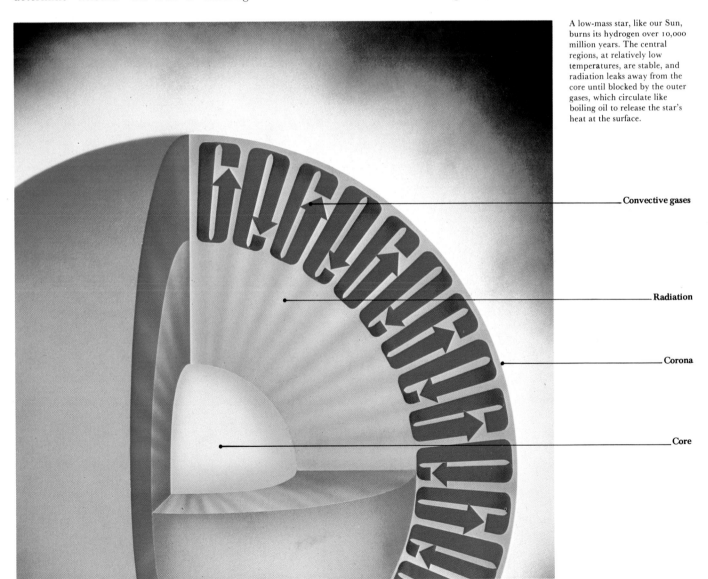

Convective gases

Radiation

Corona

Core

the main sequence depends on its mass. Though more massive stars have a greater supply of fuel, they are also very much brighter. They are profligate in their consumption of energy, and eat up their fuel stores rapidly. For example, a main-sequence star with a mass ten times that of the Sun is roughly a thousand times brighter. Consequently it is consuming its total supply of hydrogen a hundred times faster than the Sun, and it can only live for a hundredth of the Sun's life span. Contrary to expectation, it would not be wise to bet on the longevity of a massive star; choose a faint, cool, low mass star if you seek a long, stable, quiet life.

Detailed calculations show that a star like the Sun will exist as a main-sequence star for

*Right:* A false colour photograph of a solar flare licking thousands of miles out into space shows the hot-spot (white) associated with such events.

*Below:* A solar flare billows up in 1969, releasing a stream of particles that caused a severe magnetic storm on Earth when they arrived a few hours after the outburst.

# The Seething Sun

At the Sun's surface, gases carrying away heat from the interior seeth and bubble at about 6,000°C. But the normal turbulence of convection is often disturbed by magnetic storms that tear the surface, or photosphere, into sunspots, cooler regions sometimes thousands of miles across.

Associated with these are sweeping flares and prominences that flame anything up to half a million miles into space, curving back along lines of magnetic force. Some quiescent prominences are semi-permanent features. These violent events contribute to the streams of charged particles – the 'solar wind' – that drive across the Solar System.

On Earth, they cause magnetic storms and (possibly) changes in the weather. For reasons unknown, the surface disturbance, as measured by sunspot activity, peaks about every $11\frac{1}{2}$ years (see graph).

about 10,000 million years. The Sun itself has reached the half-way stage in its main-sequence life – in about 5,500 million years ·it will have burnt the last of its central hydrogen and will die. In contrast the bright, massive stars live for less than 10 million years and are dead and gone before the lower mass stars have really got started. As we mentioned previously, this differ-ence accounts for the concentration of bright, massive O and B stars in young, open clusters of stars; they cannot escape far from the cluster of their birth before they die.

So much for the majority of stars born in the main body of the Galaxy. But there are stars found elsewhere, in the Galactic suburbs, as it were. These stars are grouped in clusters that are

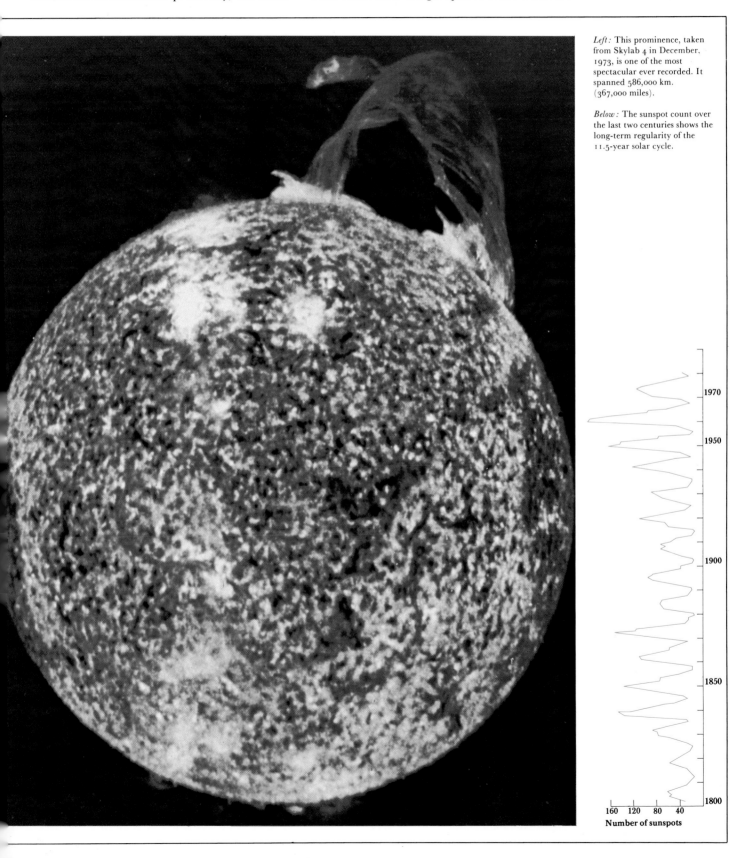

*Left:* This prominence, taken from Skylab 4 in December, 1973, is one of the most spectacular ever recorded. It spanned 586,000 km. (367,000 miles).

*Below:* The sunspot count over the last two centuries shows the long-term regularity of the 11.5-year solar cycle.

1970

1950

1900

1850

1800

160  120  80  40

**Number of sunspots**

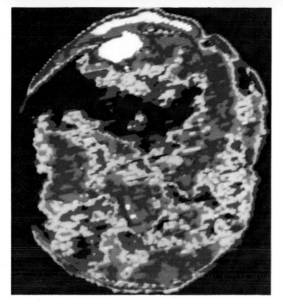
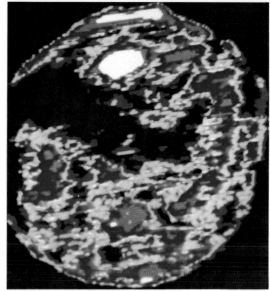

not associated with the spiral arms of the Galaxy,
or even with the galactic disc. These clusters are
spread in a sphere above and below the disc in a
swarm nearly two hundred strong. This region
surrounding the galactic disc is called the galactic
halo. Each cluster contains thousands or even
millions of stars packed in a spherical 'globule'
with a bright core containing most of the stars.

Unlike open clusters, the stars in these so-
called 'globular clusters' stay together as a
group for their whole evolutionary history.
Globular clusters contain so many stars – many
more than open clusters – that they cannot break
up. Each star is tied, or bound, by the gravitational pull of all the other stars in the cluster.

Globular clusters are not only much more
permanent groups of stars than the open clusters,
they are also much older. They were formed in the
early stages of the evolution of the Galaxy. The
Hertzsprung-Russell diagram of a globular cluster does not consist of main-sequence stars only.
Indeed, a typical Hertzsprung-Russell diagram
for a globular cluster has no stars at the top of the
main-sequence band at all. Instead the brighter
stars 'turn off' towards the right and climb in a
band up the cool side of the diagram. Instead of a
diagonal strip across the diagram, the stars form
a reversed S shape.

These cool, bright stars that have turned off
the main sequence are very large and are called
'red giants'. A typical red giant is about a hundred
times larger than the Sun. To compare a red
giant with the Sun is like comparing Concorde
with a paper dart.

The red giant state is the first step taken by a
star on the path towards its death. When all the
hydrogen is exhausted in the stellar core of a
main-sequence star, its weight is no longer supported by energy generation in the interior.

At first, hydrogen burns in a shell, rather like
an eggshell, surrounding the core of helium left
behind by the main-sequence burning stage. The
shell of burning hydrogen dumps its 'ashes' onto
this core, building it up progressively with
supplies of more and more helium. But the core,
with no energy sources of its own, cannot grow
indefinitely. When it reaches a critical mass, the
centre collapses and the outer layers of the star

expand, stretching out into space and growing in
size one hundredfold. The central regions continue trying to disappear down to a central point,
collapsing towards higher densities and increasing
in temperature, until the centre becomes hot
enough for helium to 'ignite' and undergo fusion
reactions which yield more complex elements,
principally carbon and oxygen. Just as the
collapse of the initial protostar was halted by
hydrogen fusing to helium in the centre, so the
collapse of the core is halted by helium starting
to fuse into heavier elements. Thus helium,
formed as the ashes of hydrogen burning during
the main-sequence stage, becomes the fuel for
the next chapter in a star's life, the red giant stage.

As the central regions are taken over by
helium burning, the outer regions expand in the
way we have described. The star has mush-
roomed into a huge, inflated gas-bag with a
tiny, dense nuclear burning region in the centre.

In this way, the cool red giant stars found in
globular clusters are produced. The red giants
have exhausted their central hydrogen fuel and
are on the road to senility and death.

What about the even more massive stars?
Globular clusters are groups of stars which are so
old that the most massive stars within them long
ago ceased to be even red giants. The most
massive stars have died completely, and any
remnants form a stellar graveyard of super-
dense dead stars – peculiar objects with peculiar
names: white dwarfs, neutron stars, and perhaps
even black holes.

**Senility and Death**

In massive stars the ignition of helium is only the
first in a whole series of stages in which more
complex nuclei are successively burnt. As each
nuclear fuel is burnt out and exhausted in the
centre, collapse of the inner core sets in again.
Collapse continues until the temperature is high
enough for the ashes of the previous burning
stage to ignite. Helium burns to carbon and
oxygen; carbon and oxygen burn to silicon;
silicon burns to iron.

What halts this sequence of burnings to even
more complex nuclei? At the point when iron is
being forged in the centre of a massive star, with

surrounding shells of silicon, oxygen, carbon, helium and hydrogen, the end of the road is near. When the iron core collapses there is no hope of a new reaction stepping in to save the star, for it is not possible to fuse iron into more complex elements (such as uranium) *and* release energy at the same time. Indeed, elements much more complex than iron have a tendency to undergo fission, splitting apart spontaneously into less complex elements.

The explanation of what happens to a star at this point is believed to account for the origin of elements like uranium also. The collapsing iron core is thought by astronomers to explode, with explosive nuclear reactions releasing vast amounts of energy very suddenly. This sudden outpouring of energy blows off the outermost layers of a star, and causes the brightness to increase enormously. The vast supernova explosions observed in our own and neighbouring galaxies, in which a single star becomes as bright as a whole galaxy in a couple of days, are due to a star bursting apart in this way at the end of its life. Such supernovae are the last dramatic gasp of a star in its death throes. (The name is derived from another type of stellar explosion – a 'nova', or new star – but the processes involved are different, and the two names should not be associated with each other.)

It is during supernova explosions that the creation of the more complicated elements like uranium is thought to occur. These, together with the other elements built up from hydrogen over the life of the star, are flung out into space in a vast expanding cloud of gas. The space between the stars is replenished with gas – but not the original hydrogen and helium which collapsed to form the star. Instead it is full of oxygen, nitrogen, copper, manganese, bromine, titanium, gold, silver and all the other elements which make up our world on earth. These elements went into the mixture from which the Sun and the Solar System were later formed.

Massive stars are thus the crucibles in which the bulk of the elements with which we are familiar are created. Without these massive stars the Universe would simply be a mixture of hydrogen and helium, created during the early stages of the Universe before stars or galaxies had formed at all. It is sobering to realize that almost all the elements which make up our familiar world of water, air, earth and living tissue were formed in the deep interior of distant stars. You and I, and this book you are reading, and the ink it is printed with, once went through the raging furnace in the centre of a star.

Probably the most remarkable object in our Galaxy is the remnant of such a supernova explosion. On 4 July AD 1054 Chinese astronomers noted the sudden appearance of a 'guest star', a brilliant apparition clearly visible during daylight for about a month. An enormous amount of gas was expelled in that explosion, equal to the mass of the Sun. We can now see this ejected gas as a tangle of bright pinkish filaments expanding outwards into surrounding space, called the Crab Nebula.

Associated with the filaments of expanding gas are strong magnetic fields. High energy particles spiralling around these magnetic fields produce radio, ultra-violet, X-ray and even gamma-ray radiation within the nebula.

Near the centre of the Crab Nebula lies a most unusual star. In 1967 it was discovered that what appeared to be a relatively normal bright star was behaving in a most extraordinary fashion. It flashed on and off 30 times a second like some galactic warning beacon at the centre of the remnant clouds of the supernova.

What is this pulsing stellar remnant left

## Birth of a Supernova

The Crab Nebula, the gaseous remnants of a supernova explosion, was noted by Eastern astronomers when the light from the stupendous blast first reached the Earth on July 4, 1054. The original star, at a distance of 6,300 light years, would have been inconspicuous from Earth. But the brightness of the blast was 200 million times that of the Sun, and it was visible by day for 23 days. A Chinese account records that it was as bright as Venus; 'it had pointed rays on all sides and its colour was reddish-white.'

The object was rediscovered in the late 18th century, and received its name in the 1840's from a crab-like sketch of it by the English astronomer Lord Rosse. Since then, it has assumed a unique place in astronomy as the first known supernova remnant, the first X-ray source and the first known site of a pulsar – the tiny throbbing cinder of the cataclysm.

behind by the catastrophic explosion of AD 1054? Astronomers believe this star is a neutron star, spinning 30 times a second, and flashing (like a lighthouse) every time a bright beam of light emitted from near the surface sweeps across our line of sight.

It cannot be denied that the Crab Nebula is an exciting subject for much astronomical research, but more recently, in February 1987, the astronomical community of the world was galvanized into action to make detailed observations of a supernova that could rival the Crab in terms of research opportunities.

In the early hours of the morning of 24 February 1987, Ian Shelton, a young Canadian astronomer, was working at the Las Campanas Observatory, in northern Chile. He was working with a 25 cm (10 inch) telescope, taking photographs of the region of sky containing the Large Magellanic Cloud (LMC) as a routine part of an ongoing research programme. At the end of his night's observing run, Shelton developed the plates he had exposed and on one of them he noticed a bright spot near the image of the Tarantula Nebula, otherwise known as 30 Doradus. Being quite familiar with that region of sky, he thought the spot was a fault in the photographic emulsion of the plate. Just to make sure, he went outside and took a look at the LMC with his naked eye. Sure enough, there, right next to 30 Doradus, was a bright point of light. Shelton had become the first person in some 383 years to discover a supernova that was bright enough to be seen without either telescopes or binoculars.

By comparing high quality photographs of the region around 1987A with those taken some years previously, it was possible to identify the progenitor star as Sanduleak$-69°$ 202 – a massive, twelfth magnitude, blue supergiant.

The close proximinity (170,000 light-years) and brightness of this supernova, designated 1987A, allowed astronomers to make unprecedentedly detailed observations right across the spectrum, from gamma rays all the way to radio waves. The observations revealed that this was a Type 2 supernova, which involves the sudden collapse of the stellar core and the subsequent explosion of a very massive star (heavier than eight Suns) at the end of its life. The core collapse leads to the formation of a super-dense neutron star, while the outer layers of the star are literally blown off into space at very high speed.

For completeness, it should be mentioned that a Type 1 supernova involves the explosion of a white dwarf star, which has become overburdened by gas pulled from a large companion in a close binary system. When the accumulating mass of gas reaches a certain limit, the white dwarf suddenly contracts and ignites in a thermonuclear explosion,

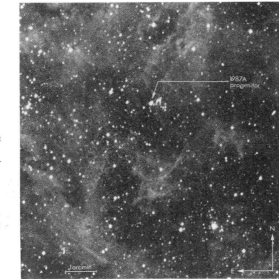

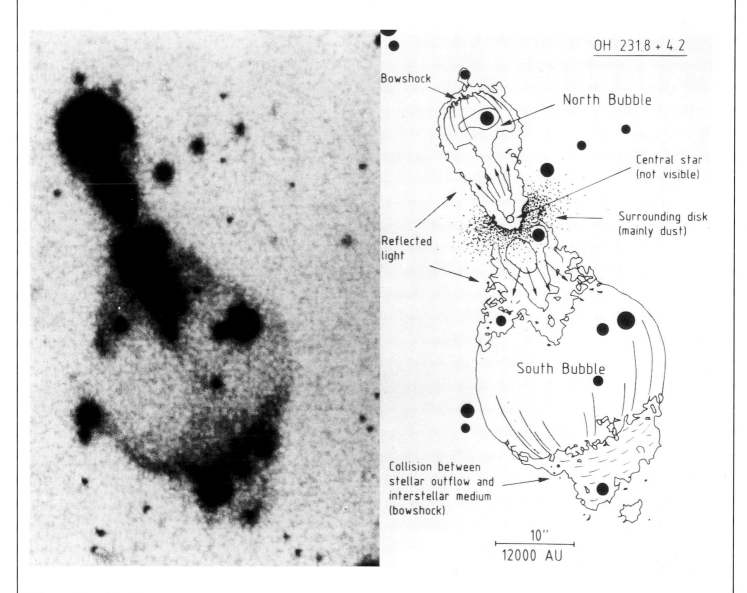

Bowshock

North Bubble

Central star
(not visible)

Surrounding disk
(mainly dust)

Reflected
light

South Bubble

Collision between
stellar outflow and
interstellar medium
(bowshock)

10"
12000 AU

# Death of a Star

All stars, no matter what type, are formed as the result of the gravitational contraction of huge clouds of interstellar gas and dust. But they do not all end their lives in the same way. Heavy stars appear to end their lives in spectacular supernova explosions, like 1987A (page 98). However, less massive stars, like our Sun, follow a somewhat different path. For example, once the Sun has exhausted its hydrogen fuel supply its diameter will increase to more than one hundred times its present size and it will become a red giant star. Thereafter it will very rapidly shed its outer layers, leaving a shrunken core surrounded by an ever-expanding, and ever thinning, shell of material. The remnant core will be a white dwarf star, which slowly cools and ultimately fades from sight. Although this scenario appears plausible to most astronomers, nobody had ever directly observed how a Sun-like star undergoes the dramatic transformation from a red giant to a white dwarf.

This crucial stage in star evolution has now been observed. The above photograph is one of the images obtained by astronomer, Bo Reipurth, of the peculiar star OH231·8+4.2 which is visible as a faint object 4000 light-

years away, in the southern constellation of Puppis. OH321·8+4.2 was previously known to be a strong radio source and an infrared emitter. Reipurth obtained red-light images of the object with a CCD camera, and discovered a most unusual structure. The images show part of the rapid transformation of a red giant star at the end of its life.

The star itself cannot be seen on the picture – it is embedded in, and obscured by, a cloud of dense material. The "bubbles" visible in the picture consist of matter which has been ejected from the star. The dark band around the "waist" is the shadow of an extremely dense disk of dust particles, orbiting the rapidly evolving central star. The material in the nebula has been ejected in directions perpendicular to the disk. The light received from the innermost part of the nebula is from the star itself, but the light from the outer part (the bubbles) is emitted from "shocked" regions where the rapidly moving stellar matter hits the surrounding interstellar gas.

The star seems to be losing mass very rapidly. Since the size of the "South Bubble" is about 0·8 light-years, and the outward velocity has been measured to be in the region

of 140 kilometres per second (about 300,000 miles per hour), the ejection must have started very recently, in astronomical terms. When this information is added to similar data for the "North Bubble", it appears that the process has been going on for only 1400 years. At the end of this short-lived transformation process, the star, which was originally rather similar to our Sun, will have lost much of its material and will have been transformed into a very small, dense object.

At some time in the future, when OH231·8+4.2 has shed a large part of its mass, the surrounding material will become less and less dense as it slowly expands. The remaining core of the star, the white dwarf, may well become visible. The entire object will then appear as a planetary nebula, consisting of a diffuse mass of gas that glows because of the irradiation by ultraviolet light from the central white dwarf. Much later, when the surrounding gas has dispersed into space, only the white dwarf will be left.

Since normal stars live for billions of years, and the mass-loss phase only lasts for a few thousand years, our chances of observing a dying star in this phase are very slim indeed.

which totally destroys it.

Astronomers will be observing 1987A, which has become the astronomical event of the decade, for many years to come, in order to follow and better understand the post-supernova evolution of massive stars.

## Stellar Remnants

Neutron stars of which the so-called 'pulsars' are special examples are one of three types of stellar remnant formed when a star dies.

The most common stellar remnants are 'white dwarfs': compact stars that are slowly cooling off in space. White dwarfs are made of strange stuff. Normal gas tends to heat up when it is squeezed. It is this response that leads to the sequence of nuclear burning reactions during stellar evolution that we have described. However, when gas is compressed to extremely high densities – such densities that a cigarette packet full of matter would weigh several tons – the electrons start to exert a special kind of pressure. They become what physicists call 'degenerate' and, as long as the star is below a critical mass, refuse to be squashed further. The electrons exert this pressure no matter what their temperature. A star composed of this material can cool without collapsing since the electron pressure can hold the star up against gravity.

Most stars end in this state, as white dwarfs the size of the earth but 100,000 times more massive. The white dwarfs cool off in space, and become steadily fainter and fainter as they slowly freeze. Eventually they form completely dead stellar stumps: worlds which are absolutely cold, and emit no radiation whatsoever.

The pressure exerted by degenerate electrons is not without limit, however. If the star is too massive, the force of gravity will overwhelm the electrons. The maximum mass of a white dwarf (first calculated by S. Chandrasekhar) is a little less than one and a half times the Sun's mass. Stars more massive than this must either lose mass before they reach the end of their life and become white dwarfs, or else they will collapse to form an even more compact star when they die. Such an object is a 'neutron star'.

If the electrons and protons are crushed together even more than they are in a white dwarf, they bind together forming neutrons. Neutrons exert their own 'neutron degeneracy' pressure which is even more powerful than that of electrons. It is able to support a star of up to three solar masses. A neutron star, being crushed even more strongly by the force of gravity, is even more compact than a white dwarf. A neutron star with the mass of the Sun would be about the size of Manhattan.

What happens if a star is squeezed even further, as must happen if the collapsing stellar remnant is more massive than three solar masses? The possibility arises, within the laws of physics as we presently understand them, that collapse continues indefinitely. All the matter showers into a central point, squashed by gravity to

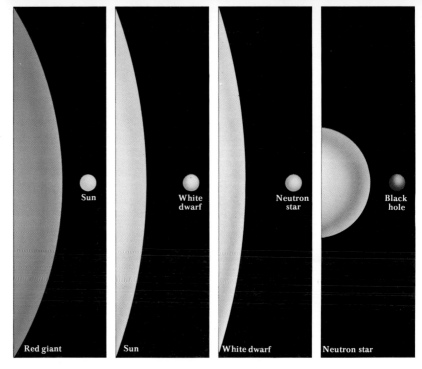

infinite density, and forms a 'black hole'. No star has ever been observed in this state, for before this condition is reached any observer will lose track of the star. It will disappear from view. As gravitational collapse beyond the neutron star state develops, densities are reached which are so enormous that light itself cannot escape from the surface. With the increasingly strong gravitational fields associated with the increasing density, the velocity needed for escape increases. Eventually a stage is reached at which this escape velocity is greater than the velocity of light. Nothing can escape, then, or ever after. The collapsing star disappears behind an impenetrable horizon – the 'event horizon' – and is hidden forever from our eyes.

How then could we 'see' a black hole if no radiation can escape from it? Only by observing the gravitational effect the matter buried inside the black hole has on material around it. As we shall see, the most likely candidate for a black hole has been found by this technique. It can be seen only indirectly, through the effect it has on a companion star in a binary star system.

So far we have been following the death pangs of a massive star. Massive stars end in a catastrophic supernova explosion, leaving behind a remnant neutron star, or a black hole, or they may even disrupt entirely. In that case only the filamentary supernova remnant expanding into the interstellar medium remains to tell the tale of a star's death.

The less massive stars die less dramatically, but form just as beautiful structures in space. Stars like the Sun never even reach the stage of burning carbon in their interior. At some point the whole outer envelope lifts off and drifts out into space. The outer region of the star becomes less and less dense as it expands outwards, and eventually the hot, central core of the star is revealed surrounded by a tenuous halo of gas like some vast, spherical, cosmic smoke ring. Such structures are observed as planetary nebulae. The central star in a planetary nebula is con-

There is a 100-million-fold difference in size between the largest and smallest types of star. This diagram compares types that are all of one solar mass. In fact, some red giants are even bigger. If Betelgeuse, for instance, were placed in our Solar System, its surface would lie beyond the orbit of Mars. White dwarfs and neutron stars have well defined limits, but black holes (the very existence of which is still controversial) could be of any size.

Star type (1 solar mass) and diameter:
*Red giant:* 140 million km./ 87 million miles.
*Sun:* 1,392,000 km./ 870,000 miles.
*White dwarf:* 13,000 km./ 8,000 miles.
*Neutron star:* 16 km./ 10 miles.
*Black hole:* 2.5 km./ 1.5 miles.

tracting to become a white dwarf, and will finally cool to become a lifeless stellar stump.

### Pulsating Stars

In addition to the red giants, the white dwarfs, the neutron stars, and the planetary nebulae, there exist a number of special types of star whose strange properties are caused by peculiarities in their structure. Among the most fascinating are the pulsating stars, stars whose surface layers rise up and sink down just as if the stars were breathing. Various classes of pulsating stars exist, including the RR Lyrae stars, the Mira variables, and the Cepheids. The Cepheids are particularly important because the scale of the Universe has been determined to a large extent using their pulsations as a yardstick.

As a Cepheid pulsates its brightness, surface temperature and radius all vary. In 1908 Henrietta Leavitt discovered that amongst the Cepheids in the Small Magellanic Cloud (a small galaxy, neighbour to our own) the brightest had the longest period of pulsation. She showed that the correlation between the period and the intrinsic brightness was so good that, by measuring the period of a Cepheid, its intrinsic, or absolute, brightness could be deduced. By observing the apparent brightness of a Cepheid as seen from Earth, and comparing this with the intrinsic brightness deduced from its period, the distance of a Cepheid may be derived. This method is exactly the same as the method one might use to judge the distance of a car from the brightness of its headlights. When applied to Cepheids the technique has proved to be one of the most important methods for determining the size and structure of our own Galaxy and the distance of nearby galaxies.

The pulsating Cepheids oscillate with typical periods of about five days, whereas the RR Lyrae stars oscillate with periods of less than a day, and the Mira variables with periods as long as several years. Each type of oscillating star sits in a particular region of the Hertzsprung-Russell diagram and each is at a particular stage of post main-sequence evolution.

Why should a star oscillate like this, bouncing in and out like the lid on a pan of boiling water? The pan lid lifts up because of the pressure of the steam trapped underneath. The steam then escapes and the lid drops down. Again the pressure builds up and the lid rises, only to drop once more when the supporting steam escapes. The lid will bounce up and down just so long as there is a source of steam, that is, as long as water is boiling in the pan.

In a similar way the outer layers of pulsating stars act as a trap on the radiation inside, damming up the photons in their flight to the surface. The barrier presented by these outer layers causes the temperature, and the pressure, to build up and the star expands. In this expanded state the radiation escapes more easily – the lid has lifted off. With more radiation escaping, the brightness of the star increases, and the outer layers cool. The surface layer, no longer hot enough to support itself, collapses and radiation is once more dammed up. Radiation, trying to escape, forces the star to oscillate at its natural frequency of vibration. The brightness, surface temperature and radius of the star all vary during these pulsations.

### Cannibalism Amongst Stars

Stars not only pulsate, they also consume one another. In the past few years cannibalism amongst stars has become recognized as quite common.

Stars can only consume one another in a binary system (a double star system in which two stars orbit around their common centre of gravity, in exactly the same way as the earth and moon orbit about one another). If the two stars in a binary system are close enough, they may run into a serious difficulty as they evolve. A star evolving off the main sequence expands and eventually becomes a red giant 100 times larger than its youthful main-sequence state. As the two stars in a binary system age, the more massive star, or primary as it is called, will burn out its supplies of central hydrogen fuel first. It will start to expand. However, if its less massive companion, or secondary, is close enough, it will start to interfere with the expansion of the more massive star. It will disturb the primary star's envelope because of the clutch of its gravitational attraction. A star is held together by the gravitational attraction of the gas *inside* pulling inwards on the gas *outside*. If the companion star is too close, the gravitational attraction of the companion will overwhelm the inward pull that holds the primary star together. The primary takes on a pear-shaped form, the point of the pear pointing towards the secondary star.

When the primary expands further the secondary's feast begins. For gas at the point of the pear is ripped off and falls in a continuous stream onto the secondary. The secondary will continue to consume its fellow star so long as it tries to evolve. The so-called Algol binary star systems

The pulse of a typical Cepheid variable star – named after the prototype in the constellation Cepheus – is related to brightness and size. Apparent brightness can therefore be used as a guide to distance. The surface, like the lid on a boiling kettle, lifts as energy builds up internally over the course of several days. Maximum brightness coincides with the bluest colour, but size, which varies by 20 per cent over the cycle, increases beyond that point as energy is dissipated. The surface then collapses again to renew the cycle.

**Colour and relative size of pulsating star**

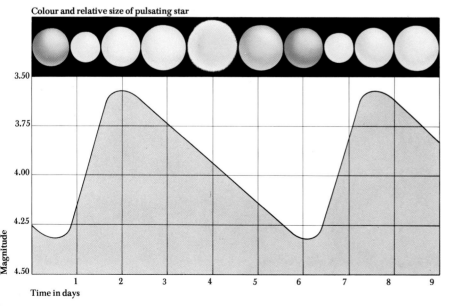

Magnitude

Time in days

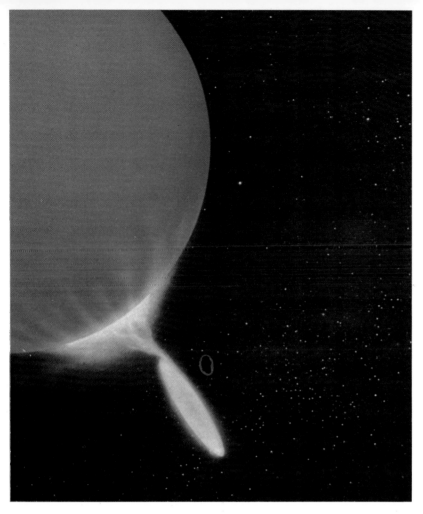

Cygnus X-1, a powerful X-ray source, is a binary – a blue supergiant 20 times the mass of the Sun and a black hole with four times the mass of the Sun which orbits its companion every few days just a million or so miles away. The enormous gravitational field of the invisible black hole (the 'hole' would be no more than two or three miles across) tears gas away from the companion. The gas is sucked into the hole as if into a whirlpool, and particles in the whirling accretion disc are heated by friction. In the process, X-rays are released.

are in this state.

Other cannibalistic binary stars are even more spectacular – they explode as novae. It has been discovered over the past 20 years that all novae are binary stars in which gas is being exchanged. In the novae the parasitic star is a white dwarf which is slowly consuming its main-sequence companion. The gas does not fall straight on the white dwarf once it is ripped off the main-sequence star. Instead it swings around the back of the white dwarf, and slowly spirals inwards onto the surface. The spiralling gas forms a disc within which the gas circulates like water spiralling around the plughole of a bath. The gas slowly falls inwards in the disc, circling faster and faster and getting hotter and hotter as it approaches the white dwarf. Finally, reaching the surface, the infalling gas is arrested. Like sparks flying from a grinding wheel, the sudden arrest of this gas liberates energy and radiation. When we look at these nova binaries we see optical radiation which is predominantly produced by this accretion process. The disc of gas is brighter than either of the two stars.

Two classes of novae exist. One class, called 'dwarf novae', erupts repeatedly in a series of small explosive outbursts. Every 30 days or so the disc grows brighter, and then slowly fades. It is thought that these irregular eruptions are due to bursts of gas falling onto the white dwarf in a series of repetitive splashes. Instead of gas passing smoothly between the two stars, the main-sequence star suffers a sequence of hiccups. A quantity of gas, about the same as that which

makes up the Earth's atmosphere, is dumped over in a sudden burst onto the disc, where it slowly spirals into the white dwarf. With each burst the disc brightens, and the binary goes through a small-scale nova eruption.

The second class of novae, called 'classical novae', is much more spectacular. In this case the white dwarf erupts, flinging a vast cloud of gas away from the whole binary system. In its brightest state the binary system cannot even be seen. It is completely enveloped in a cloud of expanding gas, flooding out at a speed close to a million miles an hour. Only as the nova fades does the cloud of gas disperse, gradually thinning out and revealing the central binary system at its centre.

These classical nova explosions are thought to be true explosions, gigantic nuclear bombs burning off on the white dwarf surface. The gas collected by the white dwarf from the companion star is hydrogen, a potential nuclear fuel. When sufficient hydrogen has collected on the surface of the white dwarf, nuclear reactions start to burn in the accreted hydrogen envelope. These fusion reactions are not controlled in the way they are in the centre of the star. They may grow rapidly, liberating energy in an explosive detonation, and eject the outer envelope of the white dwarf into space.

Finally, perhaps the most extraordinary cannibalistic stars are the recently discovered X-ray sources. These are rather like the novae, but in this case the accreting star is a neutron star rather than a white dwarf. The neutron star is so small that the infalling gas is moving extremely fast by the time it reaches the stellar surface. The gas is so hot that the radiation escaping from the disc is emitted predominantly in the X-ray region of the electromagnetic spectrum. When we observe these systems with X-ray telescopes we are seeing gas at temperatures approaching those attained in the centres of stars.

Many of these X-ray sources have fascinating properties, but probably the most intriguing of them all is called Cyg X-1. Cyg X-1 was one of the earliest X-ray sources to be discovered, in the constellation of Cygnus. In this X-ray binary the accreting star is calculated as being greater than three times the Sun's mass, which is too massive for a neutron star. Of all the objects known Cyg X-1 is thus the most likely candidate to contain a black hole, the first to be detected. Cyg X-1 may contain the most extreme possible state of a stellar remnant.

If so, many millions of years ago the black hole in Cyg X-1 was a normal star, shining through space as a result of nuclear fusion reactions in its centre. Before that it was a cloud of gas, spread out like a veil between the other stars. Now all that is left is a puzzling, massive, gravitational centre, feasting itself on its neighbouring companion, and emitting X-rays from the hot accreting gas as it falls towards, and eventually disappears behind, the black hole's impenetrable and unknowable event horizon.

# STARCLOUDS: DEATH AND REBIRTH

The space between the stars is emptier than any vacuum on Earth, but it is still rich in clouds of gas and dust. These clouds, or nebulae, are both remnants of ancient explosions and seed-beds in which new stars are born.

Some nebulae are the outer blankets of red giants of about the mass of the Sun. As heat builds up in their nuclear furnaces, such stars become unstable and throw off a smoke-ring of gas. These planetary nebulae, of which some 1,000 are known, disperse over 10,000 years, leaving their parent stars to decline into compact senility.

In other cases, two stars – a white dwarf and a large companion – interact; the white dwarf's intense gravity tears tidal-waves of gas from the companion and ejects them into space in what is known as a nova ('new star').

Two or three times a century in our Galaxy, massive stars explode in supernovae with a phenomenal release of energy, brightening to the luminosity of a billion Suns. The gas they eject – travelling at up to 10,000 km (6,000 miles) per second – plays a crucial role, for it contains many of the heavier elements familiar on Earth. In the process, the supernova stars either blast themselves to pieces or are left as tiny, dense neutron stars or black holes.

From the gas clouds of interstellar space, new stars will eventually emerge. The clouds collapse under their own gravitational influence to continue the cycle of life and death.

In this planetary nebula, the Ring, the central star glows strongly, its thermal equilibrium restored by the ejection of the gas shell.

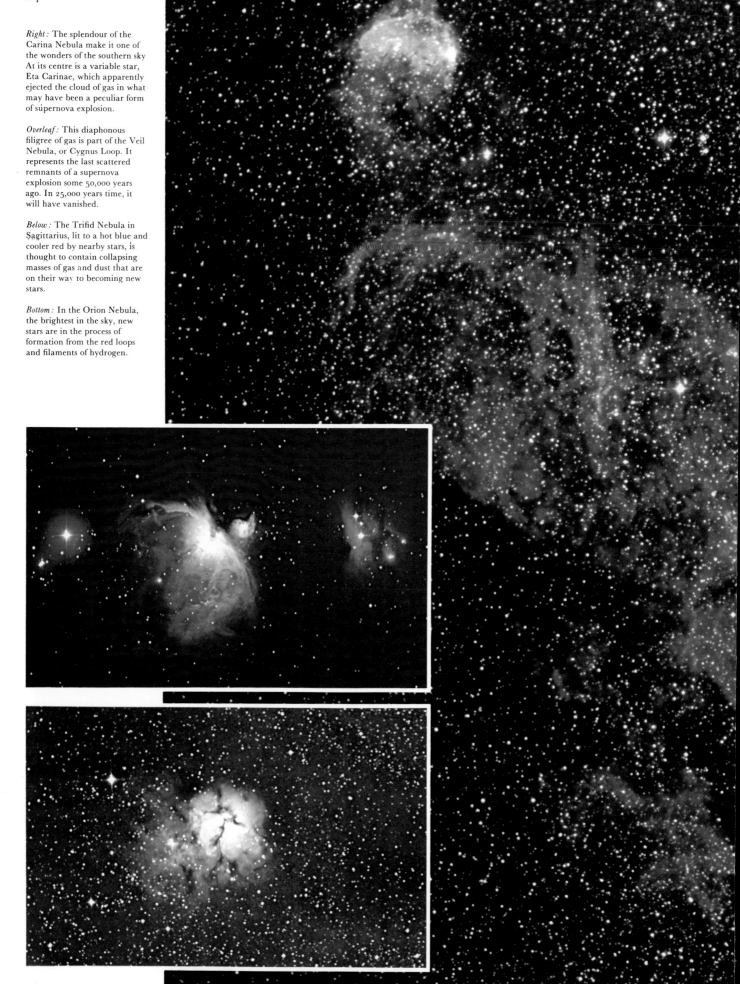

*Right:* The splendour of the Carina Nebula make it one of the wonders of the southern sky At its centre is a variable star, Eta Carinae, which apparently ejected the cloud of gas in what may have been a peculiar form of supernova explosion.

*Overleaf:* This diaphonous filigree of gas is part of the Veil Nebula, or Cygnus Loop. It represents the last scattered remnants of a supernova explosion some 50,000 years ago. In 25,000 years time, it will have vanished.

*Below:* The Trifid Nebula in Sagittarius, lit to a hot blue and cooler red by nearby stars, is thought to contain collapsing masses of gas and dust that are on their way to becoming new stars.

*Bottom:* In the Orion Nebula, the brightest in the sky, new stars are in the process of formation from the red loops and filaments of hydrogen.

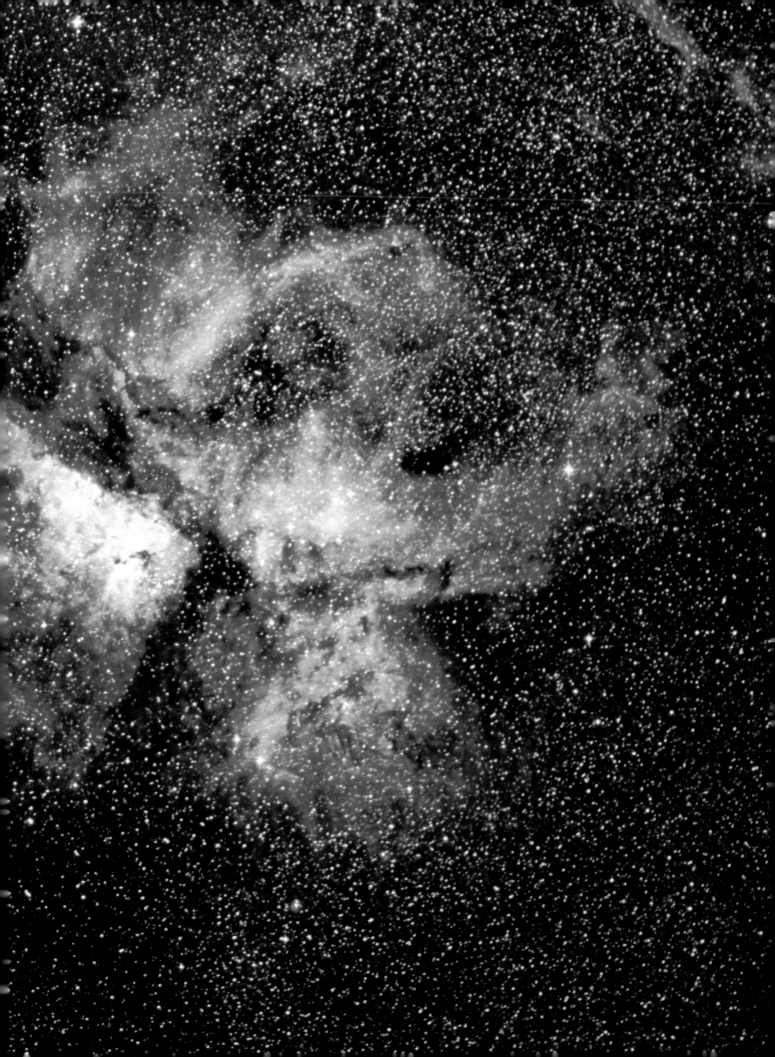

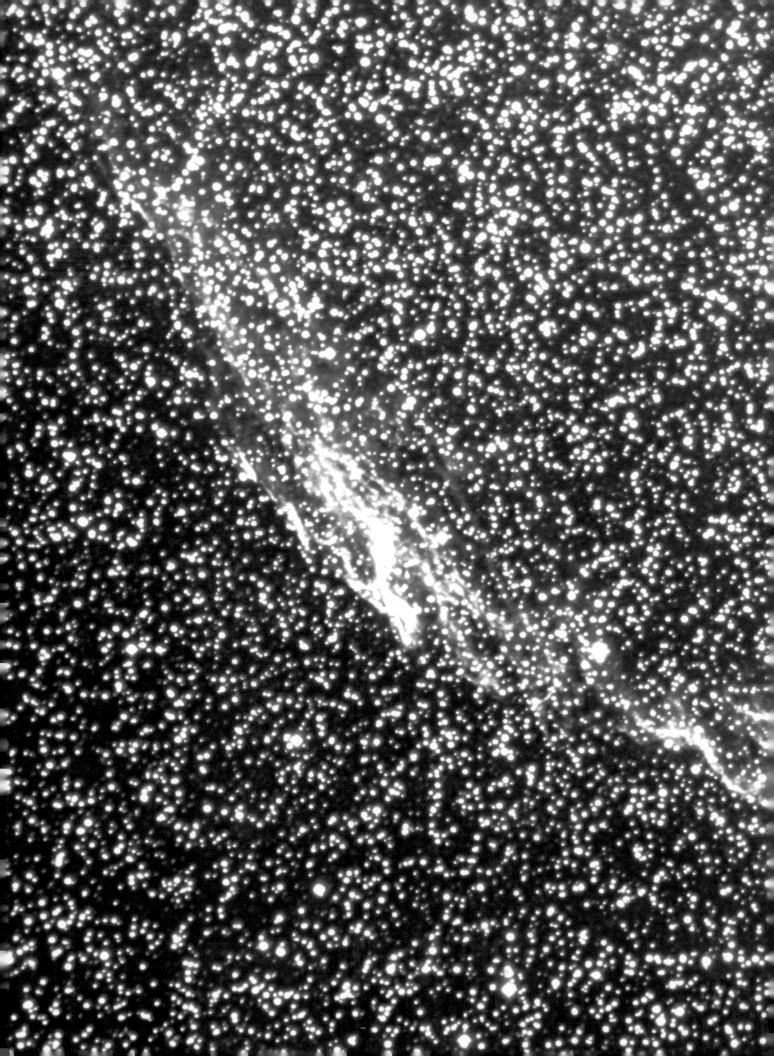

# 6/THE LIVING VOID

The possibility that other life-forms may exist in the
Universe and that we may one day contact them – has
been a staple of science-fiction novels for a century or
more. Now, however, the subject has moved out of the
realms of fiction. By asking questions about the formation
of planets, astronomers hope to learn more about the
nature of life and the chances of its recurrence
elsewhere. By searching the skies for other planets from
Earth and later with telescopes based in space (left),
astronomers hope to answer for the first time the
question: Are we alone? Intuitively, many in the past
have said that we cannot be. Perhaps we shall soon
know. News that the Earth is populated by
technologically-advanced beings has already reached the
nearest stars. If there is anyone listening, we could make
contact with an alien civilization within a generation.

On 8 April, 1960, an American radio astronomer, Dr Frank Drake, made mankind's first deliberate attempt to pick up radio messages coming from another civilization in space when he turned the 26-m. radio telescope of the National Radio Astronomy Observatory at Green Bank, West Virginia, towards the nearby stars Tau Ceti and Epsilon Eridani. These stars, 12 and 11 light years away respectively, are similar to the Sun, and thus likely candidates to have planetary systems that might support life like our own. Drake's interstellar listening attempt was whimsically named Project Ozma, after the princess of the fictional land of Oz, a place 'populated by strange and exotic beings'.

After three months of listening to Tau Ceti and Epsilon Eridani, no alien signals had been heard and the experiment was abandoned. Drake confessed that he was not particularly surprised at the result. Another civilization that close to us would be celestial overpopulation: if there were alien civilizations *that* close together, we would certainly be aware of the fact, for messages would be cluttering the interstellar wave-bands.

Project Ozma ended, but was not forgotten. It caused astronomers to consider seriously the science-fiction notion of alien intelligence and – if it exists – the problems of establishing contact. As a result many astronomers have come to accept the once-controversial possibility that life is widespread throughout space. This belief rests not upon speculations about UFO's or alleged alien visitations in times past, a subject which most scientists frankly reject, but comes from an assessment of the various factors involved in the origin and evolution of life.

The problems of making contact are rather different, and may be depressing to anyone hoping for immediate enlightenment from advanced civilizations. If there are intelligent beings circling the nearest star to the Sun, it will take eight years to say 'Hello' and receive a reply. For radio messages to be exchanged across the Galaxy would take 200,000 years. Nevertheless the possibility of life elsewhere now seems so high and the implications so startling that most astronomers think it worthwhile to search for clues to its existence, even if the civilizations concerned may have vanished long before their radio signals arrive on Earth, and even if communication (assuming the senders are still around) must be made on a rather extended time-scale.

## Life: A Universal Phenomenon?

One strong reason for believing in the widespread existence of life is the abundance of organic chemicals, both on Earth and, more surprisingly, in space.

Stanley Miller, an American biochemist, in 1953 conducted a classic experiment on the origin of life on Earth when he assembled a mixture of hydrogen, methane, ammonia and water vapour – gases that are believed to have surrounded the early Earth. This atmosphere has since been replaced by a new one made of the gases released from volcanoes and plants (but the giant planet Jupiter still retains a primitive atmosphere to this day). Theorizing that the primitive atmosphere could have been made to synthesize more complex chemicals and that a suitable power-source was readily available in the form of lightning, Miller passed an electric spark (to simulate lightning) through his artificial atmosphere. At the end of the experiment he had produced a wide range of complex organic molecules, including a number of amino-acids – molecules that are vital to life on Earth, for they form chains that make up proteins, the structural material of living things.

Since then, scientists have found that amino-acids are formed from constituents of the Earth's early atmosphere under a wide range of conditions, such as when shock-waves and ultra-violet light are used as the energy sources, artificial equivalents of thunderclaps and radiation from the Sun. With so many ways to make organic molecules, the seas of the primitive Earth must have been in part a complex chemical soup. From this soup, life on Earth arose. Remains of simple organisms called blue-green algae are known from rocks over 3,000 million years old. For well over half its existence, our Earth has been a host to life. Presumably, a similar pattern would hold for any Earth-like planet.

But life may not be simply the product of Earth-type environments. Space itself is a chemical crucible. Radio astronomers have found that in the dense clouds of gas where stars are forming there exist a number of complex molecules, many of which contain carbon. These molecules give off characteristic long-wavelength radiation which radio astronomers can detect. So far, about 40 molecules have been detected in interstellar space, including ammonia, water, formaldehyde,

In this stony meteorite, which fell near Murchison, Australia, researchers found amino acids, the building blocks of protein and essential ingredients of terrestrial life. This is direct evidence that complex organic compounds were synthesized in space, perhaps even before the formation of the planets. This startling finding has suggested the theory that the basic chemistry of life could be established in space and then 'sown' on the Earth (and, presumably, other suitable planets in the Galaxy) as part of the normal development of solar systems.

# A Solar System Full of Life

It seems to many astronomers today unlikely that Earth has a monopoly on life. The idea is not original. Here, the French 18th-century philosopher Fontenelle argues that life is common in the Universe:

'We find that all the Planets are of the same nature, all obscure Bodies, which receive no light but from the Sun, and then send it to one another; their Motions are the same, so that hitherto they are alike; and yet if we are to believe that these vast Bodies are not inhabited, I think they were made but to little Purpose; why should Nature be so partial, as to except only the Earth? But let who will say the contrary, I must believe the Planets are peopled as well as the Earth. . . .

'I cannot help thinking it would be strange that the Earth should be so well peopled, and the other Planets not inhabited at all: For do you believe we discover (as I may say) all the Inhabitants of the Earth? There are as many kinds of invisible, as visible Creatures; . . . there are an infinity of lesser Animals, which would be imperceptible without the aid of Glasses. We see with Magnifying Glasses that the least Drop of Rain Water, Vinegar, and all other Liquids, are full of little Fishes, or Serpents, which we could never have suspected there . . . Do but consider this small Leaf; why, it is a great World, inhabited by little invisible Worms, of a vast extent. What Mountains, what Abysses are there in it?

The insects on one side, know no more of their fellow creatures or the other, than you and I can tell what they are now doing at the Antipodes; does it not stand more to reason then, that a great Planet should be inhabited . . . imagine those Animals which are yet undiscovered, and add them and those which are but lately discovered, to those we have always seen, you will find the Earth swarms with Inhabitants, and that Nature has so liberally furnished it with Animals, that she is not at all concerned for our seeing above one half of them . . . Why should Nature which is fruitful to an Excess here, be so very barren in the rest of the Planets, as to produce no living thing in 'em?'

hydrogen cyanide, and others whose tongue-twisting names testify to their chemical complexity. Among these are formic acid and methylamine, which can react together to give the amino-acid glycine. Other complex molecules may await discovery in space, perhaps including amino-acids themselves.

Already amino-acids and other molecules have been found in meteorites, lumps of rock which crash to Earth like the one which landed at Murchison, Australia, in 1969. Biochemists who analysed the meteorites, using sensitive equipment originally designed for analysing Moon rock, determined that the organic chemicals had been formed in space, and were not the result of terrestrial contamination. Most recently, Professors Sir Fred Hoyle and Chandra Wickramasinghe of University College, Cardiff, have proposed that life actually originated in the dust and gas clouds of interstellar space, and was seeded onto the Earth by comets. If so, then life must be both abundant and built along similar chemical lines throughout the Universe.

This is a highly controversial theory: other scientists believe that conditions for the formation of organic molecules would have been more favourable in the rich atmosphere of the early Earth than in the gas clouds of space. But wherever life is formed, either on Earth or out between the stars, there is clearly no shortage of the right raw materials in the Galaxy.

In theory, any star could be the Sun of a life-bearing planet. The key to life is liquid water. Each star is surrounded by a so-called life zone, a kind of interplanetary green belt in which conditions are suitable for life. The Earth, for instance, lies at the heart of the Sun's green belt. Temperatures on Earth are not so hot that water boils, as has happened on Venus, nor so cold that water freezes, as on Mars.

But in fact stars like our own Sun seem particularly favourable as centres for advanced life, for a combination of reasons: they have a sufficient energy output to warm any surrounding planets, and they are long lived enough for any life on those planets to become highly evolved.

Stars smaller and fainter than the Sun live far longer, but they have much more restricted life zones around them. A red dwarf such as Barnard's star lives an astounding 100 times longer than the predicted 10,000 million years of the Sun. Yet it glows so feebly that only the very closest planets to it would be warm enough for life. Larger stars than the Sun are much hotter and have correspondingly larger life zones, but they emit searing amounts of dangerous short-wavelength radiation, and they burn out much more quickly than the Sun. For instance, the bright star Sirius is over twice as massive as the Sun, and has approximately one-tenth the predicted lifetime, which means that it can live for no more than 1,000 million years. Our Sun, which is an average star, seems to have a suitable compromise between energy output and lifetime, and so astronomers have concentrated their search for intelligent extraterrestrial life on stars of similar type to the Sun.

What are the chances of intelligent life emerging from the raw materials, the chemical soups of dust-clouds and planets? And what are the chances of the emergence, successively, of advanced civilizations and of contact between them and us? Frank Drake has originated a rough-and-ready method to help answer these questions. His approach has been discussed at various scientific meetings, most prominently in 1971 when Soviet and American specialists met for an international scientific summit on extraterrestrial life at the Byurakan Astrophysical Observatory in Soviet Armenia.

Drake devised an equation that multiplies seven factors basic to the existence of intelligent extraterrestrial life:
- the average rate of star formation in the Galaxy;
- the fraction of stars with planets;
- the number of planets suitable for life;
- the percentage of planets on which life actually does arise;
- the likelihood of intelligent life;

– the desire of that life-form to communicate;
– and, finally, the average longevity of civilizations, like our own, which satisfy all the above requirements.

The average rate of star formation over the history of the Galaxy can be easily estimated. There are at least 100,000 million stars in the Galaxy, and the Galaxy is approximately 10,000 million years old. Simple division gives an average birth-rate of 10 stars per year.

After this, however, the figures become mere guesswork. Astronomers have estimated that at least one star in ten may have planets, but the number of those planets that will be suitable for life is even less certain. In our own solar system,

only one planet is ideal for life (though Mars is nearly favourable). It seems a fair guess that other solar systems have an average of one habitable planet.

We can only guess that life is likely to arise wherever conditions are right, and that in many cases it will develop into an intelligent, technological civilization, given sufficient time. But the number of other civilizations around at present depends on how long each survives and this is the factor that has caused most disagreement among scientists. If each new group of beings in space rapidly wipes itself out, there will be very few around for us to talk to. On the other hand, if each civilization never dies out, then the Galaxy

# A Nearby Sun, A Nearby Planet

No planets of stars other than our Sun are visible to Earth-based telescopes. Yet one star – Barnard's Star, a faint red dwarf a mere 5.9 light years away in the constellation of Ophiuchus – is known to possess at least one planet. Life is at least a possibility elsewhere.

Barnard's Star, discovered by Edwin Barnard in 1916, has the fastest 'proper motion' of any star – it is moving across laterally at 55 miles a second and approaching us at 67 miles a second.

Astronomers realized the star offered a unique opportunity to test for the existence of other planets. Any large companion would be detectable from the slight irregularity imparted to the star's path as both bodies revolved around their common centre, the barycentre (far right). The rapid motion of Barnard's Star would show up any irregularities more rapidly.

The task was formidable. One man, Peter van de Kamp, of Sproul Observatory, Pennsylvania, has devoted over 20 years to it. In 1956, after 18 years of studying thousands of photographs, he identified an irregularity with a cycle of 25 years, then spent another seven years verifying the results. Seen from the Earth, the irregularity is minute – the equivalent of detecting a one-inch movement at 100 miles.

Van de Kamp concluded one planet a little larger than Jupiter or two smaller ones could be responsible for the irregularity. If one, it would have to be in a highly elliptical orbit. It seemed more likely that any planets would have nearly circular orbits, and van de Kamp suggests Barnard's Star has two planets, in orbits 4.7 and 2.8 times the Earth–Sun distance respectively and with periods of 26 and 12 years.

His theory is summarized in the diagram at right, which shows the straight-line path of the system's barycentre. Barnard's Star, its deviations exaggerated, is shown at 25-year intervals, when the planets might be expected to line up.

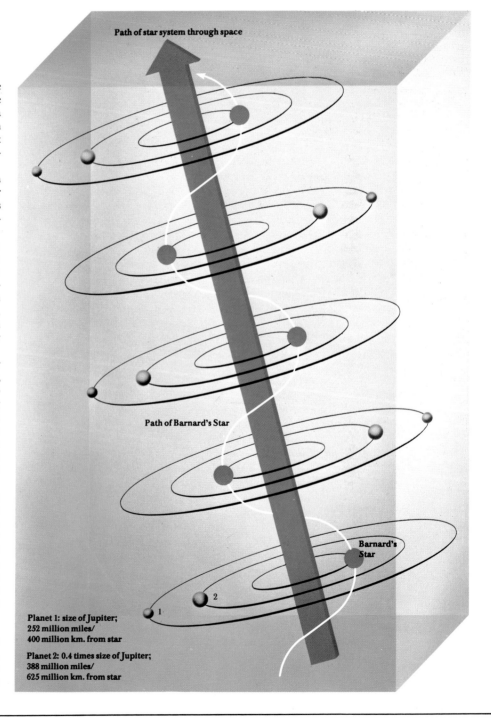

Path of star system through space

Path of Barnard's Star

Barnard's Star

Planet 1: size of Jupiter;
252 million miles/
400 million km. from star

Planet 2: 0.4 times size of Jupiter;
388 million miles/
625 million km. from star

should be literally crawling with aliens. Apparently, it is not.

The scientists at the Byurakan conference estimated that the average lifetime of galactic civilizations is 10 million years (though this seems to many a very optimistic figure, given the speed with which we have developed enough nuclear hardware to blast ourselves out of existence). Inserting this figure into the Drake equation along with plausible values for the other factors gives us an estimate of the frequency of other civilizations in space. And it turns out that there could be a total of one million civilizations like ourselves or more advanced throughout the Galaxy. This means that there could be one

These three diagrams illustrate the principle of the centre of mass, or barycentre. Any system of orbiting objects – whether our own Earth–Moon system, or a Sun and planets – orbit each other around a common centre, which is comparable to a scale's point of balance. In the case of the Earth and Moon, the barycentre is inside the Earth, and the irregularity imparted by the Moon is correspondingly minute (*below*).

civilization between every 100,000 stars, and that the nearest neighbours with whom we may communicate by radio are several hundred light years away. In other words, the chances would be that we have no near neighbours.

Some scientists naturally think this figure of one million civilizations could be too low, while others regard it as a gross overestimate. To get a better answer to the problem, scientists are seeking to refine their knowledge of the various factors in the Drake equation. Chief among these factors is the frequency of planetary systems around stars. Earlier this century, planetary systems were thought to be rare. At that time, the standard theory said that the planets of our own solar system were formed from a filament of gas pulled out of the Sun by a passing star, and such a close encounter between stars would be extremely unlikely.

But astronomers now believe that the births of stars and planets are intimately linked. According to modern theories, stars are surrounded by a ring of material left over after their formation. In many cases this material may form a second star, but in others it will give rise to a system of planets. Therefore, if a star is not a member of a binary system we should expect it to have planets.

Exactly what proportion of stars is double or multiple? Two astronomers at Kitt Peak Observatory, Arizona – Helmut Abt and Saul Levy – have examined this question. They studied 123 stars like the Sun in our stellar neighbourhood, finding that two-thirds of them were double, while the rest had no detectable stellar companions. Abt and Levy predicted that these stars also had companions that were too faint to be seen.

A true star needs a mass of at least seven per cent the mass of the Sun, otherwise conditions in its interior do not become extreme enough to spark nuclear reactions. Objects which lie between this limiting mass and the mass of Jupiter are called degenerate stars; they are nevertheless too large to be regarded as planets.

Some of the stars studied by Abt and Levy will have degenerate stellar companions. But, according to the two astronomers' calculations approximately 20% of the stars they sampled should be accompanied by genuine planets. If this figure is correct, and if it applies to all stars as well as the small sample actually studied, there must be a vast number of potential homes for life in space. The chances of our having neighbours from whom we could expect a reply within a century increase accordingly.

We cannot see planets around other stars with existing telescopes on Earth because they are too faint. But while no direct evidence exists of other planetary systems in space, there are a number of indirect methods by which planets might be detected, one of which has already produced encouraging results. This method involves measuring the slight position changes of a star as it moves through space (its *proper motion*). If the star is single, its proper motion should be in a straight line. But if it has a companion, the gravitational pull of the companion will cause the star to wobble, the amount of the wobble depending on the companion's mass and distance.

At Sproul Observatory in Pennsylvania, Professor Peter van de Kamp has been photographing the motion of Barnard's star since 1937. Barnard's star, a red dwarf six light years away, is the second closest star to the Sun. Van de Kamp has announced that the star shows a slight wobble in its motion which indicates it is accompanied by two planets similar in size to Jupiter and Saturn, orbiting the star every 12 and 26 years respectively. Smaller planets like Earth might also exist, but would produce too small an effect to be detected. Astronomers at other observatories are photographing Barnard's star and other stars in an effort to confirm van de Kamp's results.

Another method of detecting planets is to measure the slight Doppler shift in a star's light as it swings around in orbit with its companions. Very sensitive measurements are required to detect this motion, and are now being undertaken by astronomers at Kitt Peak observatory. Results are not expected for several years yet. Further in the future, it might be possible to see planets directly by special optical systems attached to telescopes in space.

The factors in the Drake equation governing the evolution of intelligence and technology are difficult to assess because we have only the example of ourselves to go on. Anthropologists can trace the dawn of man to three or four million years ago on the plains of Africa. Then, several species of ape-man lived side by side, but only one species survived to become true man. *Homo sapiens* as we know him today has existed for perhaps 40,000 years, and there have been no physical changes in mankind since that time.

We are very recent arrivals in the history of the Earth, the product of a long and complex evolutionary chain. It seems inconceivable that the same evolutionary path could be duplicated exactly anywhere else, so how can we hope to find anyone in space remotely resembling our-

# The Canals of Mars

One of the oddest controversies in the history of astronomy is that concerning the canals of Mars. The idea was first popularized by the Italian astronomer, Giovanni Schiaparelli, after observations in 1877. Schiaparelli suggested that the *canali* were systems of rivers that distributed melt-water from polar ice. His ideas were extended by Percival Lowell at Flagstaff, Arizona, who claimed the canals must be artificial, an idea that inspired countless science fiction stories until the Viking missions revealed Mars's dusty, rock-strewn and cratered surface. Schiaparelli and Lowell's work, which reflects the difficulties of observing accurately through the Earth's turbulent atmosphere, was purely wishful thinking. The vague, shadowy patches were joined in the mind's eye, as beads seen from a distance seem to form single lines.

It is startling now to see how highly developed their theories were and how little supported by the reality of the Martian surface. As Schiaparelli wrote:

'The polar snows of Mars prove in an incontrovertible manner, that this planet, like the Earth, is surrounded by an atmosphere capable of transporting vapor from one place to another. These snows are in fact precipitations of vapor, condensed by the cold, and carried with it successively. How carried with it, if not by atmospheric movement? . . .

'All the vast extent of the continents is furrowed upon every side by a network of numerous lines or fine stripes of a more or less pronounced dark colour, whose aspect is very variable. These traverse the planet for long distances in regular lines, that do not at all resemble the winding courses of our streams.

Some of the shorter ones do not reach 500 kilometers (300 miles), others on the other hand extend for many thousands, occupying a quarter or sometimes even a third of a circumference of the planet . . .

'As far as we have been able to observe them hitherto, they are certainly fixed configurations upon the planet. The Nilosyrtis [a canal Schiaparelli estimated at 200–300 km. (120–180 miles) wide] has been seen in that place for nearly one hundred years, and some of the others for at least thirty years. Their length and arrangement are constant, or vary only between very narrow limits. Each of them always begins and ends between the same regions. But their appearance and their degree of visibility vary greatly, for all of them, from one opposition to another, and even from one week to another . . .

'Every canal (for now we shall so call them), opens at its ends either into a sea, or into a lake, or into another canal, or else into the intersection of several other canals. None of them have yet been seen cut off in the middle of the continent, remaining without beginning or without end. This fact is of the highest importance.

'The canals may intersect among themselves at all possible angles, but by preference they converge towards the small spots to which we have given the name of lakes . . . That the lines called canals are truly great furrows or depressions in the surface of the planet, destined for the passage of the liquid mass, and constituting for it a true hydrographic system, is demonstrated by the phenomena which are observed during the melting of the northern snows . . . Such a state of things does not cease, until the snow,

reduced to its minimum area, ceases to melt. Then the breadth of the canals diminishes, the temporary sea disappears, and the yellow region again returns to its former area . . . We conclude, therefore, that the canals are such in fact, and not only in name. The network formed by these was probably determined in its origin in the geological state of the planet, and has come to be slowly elaborated in the course of centuries. It is not necessary to suppose them the work of intelligent beings.'

Upon which theory Lowell built his own:

'To review, now, the chain of reasoning by which we have been led to regard it probable that upon the surface of Mars we see the effects of local intelligence. We find, in the first place, that the broad physical conditions of the planet are not antagonistic to some form of life; secondly, that there is an apparent dearth of water upon the planet's surface, and therefore, if beings of sufficient intelligence inhabited it, they would have to resort to irrigation to support life; thirdly, that there turns out to be a network of markings covering the disk precisely counterparting what a system of irrigation would look like; and, lastly, that there is a set of spots placed where we should expect to find the lands thus artificially fertilized, and behaving as such constructed oases should . . .

The fundamental fact in the matter is the dearth of water. If we keep this in mind, we shall see that many of the objections that spontaneously arise answer themselves. The supposed Herculean task of constructing such canals disappears at once; for, if the canals be dug for irrigation purposes, it is evident that what we see, and call by ellipsis the canal, is

selves? Fortunately, evolution has the knack of reaching the same result by several different pathways. For instance, several different types of creature have independently invented flight: insects, mammals, reptiles, and even fish. Therefore there seem to be good reasons to suppose that intelligent, technological beings similar to ourselves have arisen on many planets around solar-type stars throughout the Galaxy.

Of course, local conditions will impose certain differences. On a high-gravity planet, for example, creatures (if they had muscles and bones like ours) are likely to be squat and heavy; whereas on a low-gravity planet the beings might be tall and slender, with large noses to breathe the thin air. On planets with vast expanses of flat land, some creatures may have developed wheels rather than legs. If the planet is cold, they could be hairy, and possibly white like polar bears. Some aliens may even look like centaurs, with four legs and two arms.

Such speculations go beyond the strict numerical predictions of the Drake equation. But what that equation does show is that the chances of other life existing in space are sufficiently good for it to be worth our while making the effort to look.

## The Search for Life Nearby

Beginning on our own doorstep, are there any signs of life in the solar system?

Mercury, the closest planet to the Sun, is an airless and waterless body of extreme temperatures. Life as we know it could not possibly exist there. Venus is not much better. It has a dense atmosphere, but this traps the Sun's heat to produce roasting temperatures, and the atmosphere at the surface bears down with 90 times the pressure of the atmosphere on Earth. Suitable conditions for life may exist higher in the atmosphere, but it's a slim hope.

We know that the Moon is lifeless, having sampled its rocks, but Mars has always looked a likely home of life. At the turn of the century the American astronomer Percival Lowell drew a whole network of so-called canals crossing the planet's surface. He believed that the canals

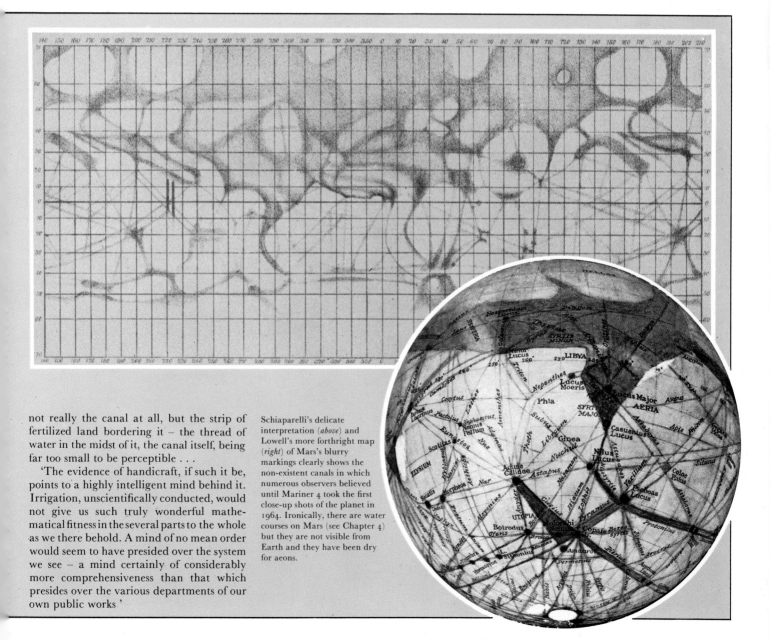

not really the canal at all, but the strip of fertilized land bordering it – the thread of water in the midst of it, the canal itself, being far too small to be perceptible . . .

'The evidence of handicraft, if such it be, points to a highly intelligent mind behind it. Irrigation, unscientifically conducted, would not give us such truly wonderful mathematical fitness in the several parts to the whole as we there behold. A mind of no mean order would seem to have presided over the system we see – a mind certainly of considerably more comprehensiveness than that which presides over the various departments of our own public works '

Schiaparelli's delicate interpretation (*above*) and Lowell's more forthright map (*right*) of Mars's blurry markings clearly shows the non-existent canals in which numerous observers believed until Mariner 4 took the first close-up shots of the planet in 1964. Ironically, there are water courses on Mars (see Chapter 4) but they are not visible from Earth and they have been dry for aeons.

were dug by Martians to bring melt water from the planet's polar caps to their crops near the equator, and he wrote books describing his vision of an inhabited Mars.

Other astronomers failed to see the canals he so confidently drew, but that didn't prevent the idea of an inhabited Mars catching on in the public imagination and in science fiction. While Lowell's ideas of Martian civilization were officially frowned upon, many astronomers were at least prepared to concede that the dark areas visible on Mars could be areas of vegetation, such as mosses or lichens, and this view was held into the 1960's.

Then, space probes began to change our view of Mars. Mariner 4 in 1965 showed that Mars is more hostile than had previously been supposed, with thin air and a crater-strewn surface. Of large expanses of vegetation there were no signs.

This gloomy picture of Mars persisted until 1971, when Mariner 9 made a complete photographic map of Mars from orbit and ignited new hope among the life-seekers. Most important among the features spied by Mariner 9 were what appeared to be dried-up river channels, as though water had once flowed freely across the surface of Mars, possibly in times of more clement climate. Giant volcanoes were discovered which might in the past have erupted sufficient gases

to form a dense atmosphere, along with liquid water. Therefore some kind of micro-organisms might cling to existence in the soil of Mars. The only way to find out was to go down and look.

In 1976, two Viking spacecraft were sent to search for life on Mars. Each of the identical craft carried twin cameras for photographing the surface, and each was equipped with an on-board biological laboratory to analyse samples of soil. A long mechanical arm reached out to scoop up soil for analysis.

Before the Vikings landed on Mars, optimistic biologists had speculated that organisms large enough to be visible to the cameras might exist on the surface. These organisms, the speculation ran, would obtain their water from the rocks or from thin nightly frosts, and would be protected from the Sun's ultra-violet radiation by parasols made of silica.

Both Vikings were targeted to lowland areas on Mars. One probe landed in a basin called Chryse where water was believed to have flowed during the wet times on Mars, and the second made landfall in an area called Utopia, which is covered with frosts in winter. Colour photographs from the Viking landers showed that the surface of Mars is a red, stony desert. Dust blown into the thin atmosphere makes the sky pink. Despite careful scrutiny, not a single cactus nor the

The Viking biology experiments, packed into a miniature laboratory, all recorded a number of reactions, all unexpected. But space scientists have now concluded that the results can be explained as chemical and not biological reactions.

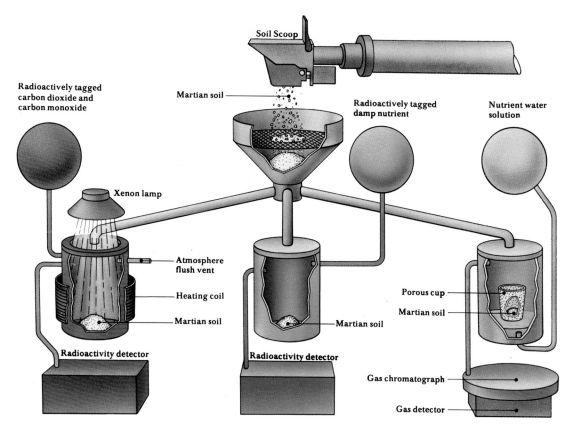

Soil Scoop

Radioactively tagged
carbon dioxide and
carbon monoxide

Martian soil

Radioactively tagged
damp nutrient

Nutrient water
solution

Xenon lamp

Atmosphere
flush vent

Heating coil

Martian soil

Martian soil

Porous cup

Martian soil

Radioactivity detector

Radioactivity detector

Gas chromatograph

Gas detector

The pyrolytic release experiment was designed to detect photosynthesis, which would reveal the existence of plants, algae or bacteria. Radio-active carbon was introduced to a sample of soil. This was incubated and heated and the gases given off measured to see whether any carbon had been assimilated.

The labelled release experiment assumed that any Martian micro-organisms would require water. A number of nutrients – eg. formate and lactate – labelled with radio-active carbon were introduced with water to the sample. Any micro-organisms present would ingest the nutrients and release the radio-active carbon.

The gas exchange experiment also assumed that Martian bio-chemical reactions were based on water. It was designed to measure an exchange of chemicals between any organisms and the atmosphere. The sample was placed in a porous cup and a nutrient-rich water solution, introduced first round the cup and then directly into it. Gases given off could then be measured by the chromatograph.

smallest patch of vegetation was visible. Not even a single cockroach or a sandfly hopped through the pictures. In the two locations where the Vikings landed, Mars looks depressingly dead.

Some idea of the hostile conditions on Mars was given by the Viking meteorology instruments. Even on a warm summer's afternoon, the air at the Chryse site reached $-29°C$, and at dawn had fallen to $-85°C$. At neither Viking site does the ground become warmer than $0°C$ at any time of the year. Combined with an atmospheric pressure similar to that at a height of 32 km. (20 miles) above Earth, Mars is indeed a harsh place.

Better news for the life-seekers came from the instrument that analysed the Martian atmosphere. In addition to carbon dioxide it detected a few per cent of nitrogen, the existence of which was previously unknown on Mars. Therefore the three main ingredients for life have been found on Mars: carbon, nitrogen and water. Viking's instruments also confirmed that the atmosphere of Mars was denser in the past. So biologists could hope that life had once arisen on Mars and still clung to existence in the form of micro-organisms in the soil.

Viking carried three biological experiments to find out. Together, they could detect traces of life in soil many times more barren than Earth's most barren deserts. Each experiment incubated the soil in a different way in an attempt to make Martian micro-life grow. A number of samples were scooped up by the craft's mechanical arm and tipped into the biology laboratory for analysis.

First, let's look at the 'gas-exchange experiment'. This fed a sample of Martian soil with a rich nutrient broth, and analysed the gases given off. When the experiment was run at each Viking site, puffs of carbon dioxide and oxygen were released. Experimenters concluded, though, that chemical processes in the soil were responsible, not Martian bugs. (Water vapour in the nutrient solution displaced the carbon dioxide from the dry Martian soil, and chemical reactions with oxides in the soil released the oxygen.)

The other experiments were not so easy to interpret. In the 'labelled-release experiment', a nutrient containing radioactive carbon was injected into the samples. If organisms existed in the soil, they might be expected to give themselves away by releasing gas containing the radio-actively labelled carbon. And indeed, sudden surges of radioactive gas were recorded when this experiment was run at both Viking sites. Radio-active gases were not emitted from soil samples that had been sterilized at $160°C$, while certain amounts of gas were emitted by soil that had been heated to $50°C$.

Results like this in terrestrial soils would be a clear indication of life, but Mars is a very different world. Any life there experiences continual sub-zero temperatures, and heating to $50°C$ would be expected to have a more adverse effect than actually observed. Once again, it could not be ruled out that chemical reactions with the highly oxidized soil were responsible for the results of the labelled-release experiment.

Third of the biology tests was the 'pyrolitic-release experiment'. This worked in the opposite way to the labelled-release experiment, by seeing if carbon was taken up by the soil sample. Each sample was incubated in a test chamber with an artificial atmosphere containing carbon dioxide and carbon monoxide labelled with radioactive carbon. After incubation, the sample was heated (pyrolysed) to drive off any carbon that had been assimilated, which was then measured.

Several soil samples were tested at each Viking landing site, with positive results. Something in the soil was combining with radioactively labelled carbon from the atmosphere. But the uptake of carbon was unaffected by heat treating the soil at $90°C$, and even sterilization to $175°C$ did not completely eliminate the reaction. The experimenters concluded that biological processes could not be responsible, and that soil chemistry must be the answer.

To interpret the biology experiments we must also take into account the results from a related instrument, known, somewhat forbiddingly, as the 'gas chromatograph mass spectrometer' (GCMS). This device searched for organic compounds by heating soil samples and analysing the gases driven off. It would reveal the existence of living things even if they had failed to respond to the biology experiments. To most people's surprise, there was no sign of any organic compounds in the soil of Mars, not even those molecules detected in meteorites on Earth.

This negative result is strong evidence against the possibility of life on Mars. Although the GCMS was not sensitive enough to detect a mere scattering of Martian micro-organisms in the soil sample, there should have been plenty of organic material in the form of food and dead bodies for it to register. Without a detectable level of organic molecules in the soil, it is difficult to attribute the results of any of the Viking biology experiments to life.

The possibility of Martian organisms cannot yet be completely ruled out. Some scientists believe that life may exist on Mars in certain favoured oases, or a small population of Martian organisms might exist that did not respond to the Viking experiments. Whatever the case, a more detailed examination of Mars is required to settle the matter finally. (Some plans for the future exploration of Mars are discussed in the concluding chapter.)

Jupiter, the giant planet of the solar system, appears totally unlike the Earth, yet it may be the most promising other site for life in the solar system. Its atmosphere is made of the same gases which surrounded the primitive Earth, and from which life is believed to have arisen: hydrogen, ammonia, methane, carbon dioxide and water vapour.

Organic molecules are formed in abundance from simulations of the atmosphere of Jupiter, as in simulations of the atmosphere of the primitive Earth. Typical of the molecules formed are reddish-brown tarry substances. The red and brown colouring in the swirling clouds of Jupiter is probably caused by organic molecules.

Jupiter has no solid surface beneath its clouds; instead, the gases of which it is made keep on getting denser. At its cloud tops, the temperature of Jupiter is $-120°C$, but deeper down in its atmosphere Jupiter is warmer. Liquid ammonia exists below the visible clouds of Jupiter, and this region of Jupiter's atmosphere could be a likely habitat for life using liquid ammonia as a solvent instead of water. Deeper still, about 80 km. (50 miles) below the visible cloud tops, temperatures have risen sufficiently for clouds of water vapour to exist, making this an even more favourable locale for Jovian life.

Some scientists have speculated that living organisms may float in the clouds of Jupiter, in similar fashion to fish floating in the seas of Earth. These creatures would be gas bags, filled with helium, and moving around by expelling gas. They would feed off organic molecules in the clouds around them. Possibly the Jovian organisms may themselves be coloured red. Their favoured site may be the great red spot, an updraught of warm air from deep within Jupiter. Probes diving into the atmosphere of Jupiter will be needed to discover whether or not there is life there.

Saturn is a similar planet to Jupiter except that its atmosphere is less colourful, and so there is less chance that it contains complex organic molecules. Saturn's largest satellite, Titan, is a fascinating body. With a diameter of 5,800 km. (3,625 miles), Titan is larger than our own Moon, and has a denser atmosphere than Mars. Titan may be the most favourable site for life in the outer solar system.

Beyond Saturn the solar system becomes icily cold. Uranus and Neptune are so remote that sunlight reaching them is too weak to nurture living organisms. There seems no hope of finding life that far from the Sun.

## Listening for messages from the stars

Whether or not there is life on Mars, Jupiter, or any other planet of the solar system, we will find advanced beings only by looking much further afield, beyond our own solar system. As yet, we cannot send probes to other planetary systems, but we do have the ability to exchange radio messages with beings around other stars (though such an exchange would take a decade or two for close neighbours and many centuries for more distant civilizations).

It was this realization that sparked Frank Drake's Project Ozma to pick up interstellar messages in 1960. Improvements in electronics since then have meant that current radio telescopes are far more sensitive. The world's largest individual radio astronomy dish, the 305 m radio telescope at Arecibo in Puerto Rico, is powerful enough to exchange messages with an instrument of similiar size anywhere in the Galaxy.

The first task is to find out if anyone is transmitting. Two main problems hamper the search for radio messages from the stars: where to look, and on which wavelength. If we assume that Sun-like stars are the most favourable for the emergence of advanced life, then we can easily assemble a list of nearby targets – although if we wish to look further afield than a few hundred light years the number of stars increases so rapidly that it becomes best to scan the whole sky.

Choice of wavelengths is much more difficult. In 1959, as plans for Project Ozma were being hatched, two American physicists, Giuseppe Cocconi and Philip Morrison, proposed that the natural channel for interstellar communication would be 21 centimetres. This is the wavelength emitted naturally by hydrogen gas in space, to which radio astronomers all over the Galaxy would – it could reasonably be expected – already be tuned. Project Ozma listened for artificial signals from aliens at 21 cm. wavelength, as have many other searches since then.

But other scientists engaged in the search for extra-terrestrial intelligence (SETI) have argued that the existing noise from hydrogen makes 21 centimetres the last wavelength to choose for interstellar signalling. Discoveries of emissions at specific wavelengths from various other molecules in space have complicated the issue, and radio astronomers are now abandoning the idea of a 'preferred' wavelength for interstellar communication. They think we will need to scan the entire radio window observable from Earth, which adds greatly to the search effort.

A number of searches have been made to date, concentrating on specific wavelengths. First to follow the lead of Project Ozma was a Soviet team led by Vsevolod Troitsky of Gorky State University, who in 1968 began using a 15-m. radio dish to listen to 11 nearby stars, plus the Andromeda galaxy, at a wavelength of 30 cm. Troitsky and his colleague Nikolai Kardashev subsequently set up across the Soviet Union a network of low-sensitivity aerials which remains in operation in the hope of receiving intensely powerful transmissions from advanced civilizations in the Galaxy.

In 1972, Gerrit Verschuur at the National Radio Astronomy Observatory, Green Bank, reckoned the time was ripe for a sequel to Project Ozma. He used the observatory's giant 91-m. and 43-m. radio telescopes to listen on 21 cm. to ten nearby stars, including the original Ozma stars, Tau Ceti and Epsilon Eridani. Such was the improvement in equipment that Verschuur calculated his search was over 1,000 times as sensitive as Project Ozma. But still nothing was heard.

A more comprehensive search has been made at Green Bank by Ben Zuckerman and Patrick Palmer. Again using the 91-m. and 43-m. telescopes at 21 cm. wavelength, Zuckerman and Palmer surveyed more than 600 Sun-like stars out to a distance of 80 light years. A 40 megawatt transmitter beaming through a 100-m. dish around any of those stars (equivalent to terrestrial capabilities) would have been detected. Nothing positive was heard.

Following these pilot schemes, the search for extra-terrestrial intelligence is widening out. At Ohio State University Radio Observatory, Dr Robert S. Dixon has been scanning the entire sky

at 21 cm. wavelength since 1973, and plans to continue indefinitely. At the Algonquin Radio Observatory in Canada, Paul Feldman and Alan Bridle are engaged in surveying several hundred stars at a wavelength of 1.35 cm., which is the wavelength emitted by water molecules – and might thus be a natural-seeming communications channel for water-based beings.

Frank Drake and Carl Sagan are using the Arecibo radio dish to survey several nearby galaxies in the hope of finding some super-civilization capable of transmitting over inter-galactic distances. A number of other small-scale searches have also been undertaken at Arecibo, Green Bank and elsewhere.

There have been several false alarms in the search for radio signals from space. In 1965, Soviet radio astronomers thought they had detected signals from a super-civilization deep in space. But subsequent investigation showed that they were picking up natural variable radio emission from a quasar, CTA-102. Radio astronomers at Cambridge got a bigger scare in 1967 when they found the first of the regularly ticking radio sources known as pulsars. They wondered if these might be pulsed communications from other civilizations, until it became clear that the signals were of natural origin.

Soviet radio astronomers made another false announcement in 1973 when they claimed to have picked up transmissions from an alien probe in the solar system. They subsequently realized that the probe was an American one, spying on the Soviet Union. In the Zuckerman and Palmer search, ten stars showed so-called 'glitches' or unexplained spikes of energy. These have been attributed to terrestrial interference, as none of the glitches reappeared when the stars were re-surveyed.

What sort of signal should we be on the look-out for? Probably the signals will be confined to a very narrow-frequency range, perhaps only one Hertz (cycle per second) or less. (For a given transmitter power, a narrow-frequency signal goes further than a broad-band one.)

Most normal objects in space emit over a wide range of frequencies. Even the 21 cm. radiation of hydrogen is smeared over a certain frequency range by the internal motions of the hydrogen clouds. Consequently radio astronomy receivers are tuned to accept a wide-frequency band, usually several kiloHertz, which degrades their ability to detect narrow-band signals. Therefore standard radio-astronomy receivers are far from ideal for SETI purposes.

To undertake a comprehensive search of the radio window for possible messages from the stars, we need to build special multi-channel receivers capable of examining a wide range of frequencies simultaneously with high sensitivity. And this is just what NASA plans to do in a major SETI endeavour due to begin in the early 1980's.

NASA scientists have designed a device known as a multi-channel spectral analyser, which can examine a million radio frequencies at a time.

Along with a device for decoding any signals that may be received, this will be attached to radio telescopes in the U.S. and elsewhere for a five-year search for messages from the stars.

Two groups are engaged in the project: the Ames Research Center, and the Jet Propulsion Laboratory, both in California. The JPL group plan to scan the entire sky visible to them using their 26-m. dish at Goldstone in the Mojave desert usually used for tracking space probes. Their receivers will cover all the so-called radio window from 30 cm. to 1.2 cm., which is the range of radio wavelengths reaching the surface of the Earth. Longer wavelengths than 30 cm. are swamped by background noise from the Galaxy, while wavelengths shorter than 1.2 cm. are hampered by noise from the Earth's atmosphere.

How will we recognize that first interstellar signal and what will it say? Radio noise confined to a narrow-frequency band should by itself be a giveaway, but nature has played tricks on us before, and it will be necessary to examine suspicious signals carefully before announcing that they are the products of alien intelligence rather than natural emission. An interstellar 'Hello' may be a simple, continuous tone, or more likely it may come as a string of on-off pulses – a kind of celestial Morse code, like the binary code used by computers. Evidence of a message coded on the radio signal would clinch the case for an intelligent origin.

What the message might say is limited only by the imagination of the senders, and our ingenuity in decoding it. Simple mathematics is one suggested possibility, perhaps including a translation programme that will enable us to decode more complex messages to follow. But the most favoured idea is a form of interstellar television, in which the message pulses can be arranged to give a dot picture showing the aliens, their home planet, and such simple scientific facts as their biochemical make-up and aspects of their lives and technology. Surprising amounts of information can be conveyed pictorially, and this form of communication avoids all language problems.

Though most scientists believe that radio is the best way to communicate over interstellar distances (because less energy is needed to send a radio message than any other form of signal), radio is not the only way for interstellar civilizations to communicate. Laser beams at infra-red, optical and ultra-violet wavelengths have been suggested, and the Copernicus satellite has been used to search for ultra-violet laser emissions from the vicinity of the stars Tau Ceti, Epsilon Eridani, and Epsilon Indi, though without success.

Roland Bracewell, an American radio astronomer, has gone as far as to suggest that the best way of establishing communication between civilizations 100 light years or so apart is by sending automated probes to the target stars. The probes would take many centuries to get to their destinations, but once there they would orbit the stars indefinitely, awaiting the radio noise which would signal the emergence of technological civilization. The probes could then converse with them while messages announcing the discoveries were on their way to the parent star. Such a strategy cuts out the otherwise inevitable delay in establishing communication over vast distances in space, and Bracewell urges radio astronomers to be on the alert for strange radio signals coming from within our own solar system which might actually be the communication attempts of an alien probe.

## Messages from Earth

To look at the problem from the other side, what could another civilization learn of us?

Britain's Astronomer Royal, Professor Sir Martin Ryle, has been among those who have urged that we should keep our existence quiet for fear of attracting the attention of potentially hostile extraterrestrials. But already it is too late. For the past 30 years or so, high-frequency radiation from TV and radar has been streaming through the ionosphere into space, and it will inevitably be picked up by anyone listening nearby.

Most of the radio noise we are emitting today comes from UHF television stations in the United States and western Europe. Extra-terrestrials of a similar technological level to ourselves could pick these transmissions up at a distance of 25 light years, a distance encompassing about 300 stars, and would receive beams from our ballistic missile early warning radars at up to 250 light years, a range encompassing several hundred thousand stars.

Inhabitants of Alpha Centauri and beyond would not be glued to their screens for weekly episodes of *Star Trek* because programme information itself would be too weak to decipher over interstellar distances. But the aliens would be able to pick up the powerful carrier waves on which the programmes ride.

Careful analysis of Doppler shifts and other characteristics of the signals would reveal the size of the Earth, its orbit around the Sun, and other information we might expect to extract from leakage noise from another civilization that Cyclops or a large space antenna could pick up.

One short radio message has been sent to the stars. On 16 October 1974, the Arecibo radio telescope transmitted a three-minute burst of pulses towards the globular cluster M 13 in the constellation of Hercules. If the 1,679 pulses are arranged into a grid 23 characters wide by 73 deep, they form a pictogram telling the aliens about ourselves. But even if there is someone among the 300,000 stars of M 13 who will pick up the message and decode it, the cluster is 24,000 light years away so an answer cannot be expected until around AD 50,000.

Our radio messages may never have listeners, but we have sent messages of other kinds across space and time that may, millennia hence, be picked up. First to leave were small-engraved

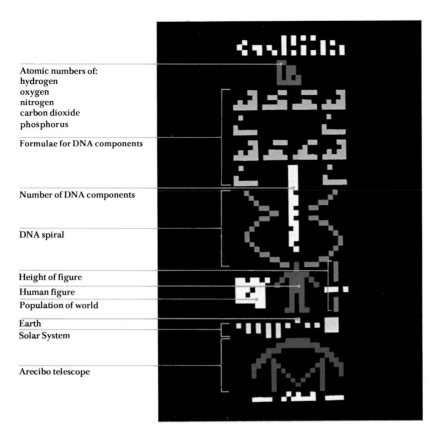

Atomic numbers of:
hydrogen
oxygen
nitrogen
carbon dioxide
phosphorus

Formulae for DNA components

Number of DNA components

DNA spiral

Height of figure
Human figure
Population of world

Earth
Solar System

Arecibo telescope

This pattern represents a string of 1,679 on-off pulses, which in 1974 were beamed by Puerto Rico's vast Arecibo radio-telescope towards the globular cluster M.13 in the constellation of Hercules. When arranged in a grid 23 × 73, the pulses form a pictorial message of numbers, atoms basic to life, and formulas for DNA (the molecule which encodes genetic instructions). Curved lines, representing DNA's spiral structure, point out a human shape. Other information – Earth's population, human height, the Solar System with the Earth displaced towards the human figure, the Arecibo telescope – would perhaps be baffling. But at least a recipient civilization would, if it could break the code, know there was intelligent life elsewhere.

plaques attached to each of the Pioneer 10 and 11 probes which will eventually exit the solar system and drift outwards into the Galaxy where some advanced civilization with an interstellar early warning radar might detect and recover them. If aliens do find the Pioneers, they will be able to tell from the plaques something about the probe's makers and their location.

A second pair of probes, called Voyager, are following the Pioneers deep into the solar system. After their planet-probing missions are over they, too, will end up floating dead and dark between the stars. They carry a different form of message in a bottle for other civilizations to decode: long-playing records. Records were chosen because large amounts of data can be squeezed onto them.

Each record begins with 115 pictures which are encoded electronically into the grooves. These pictures include views of the solar system, the Earth and its inhabitants, and human technology. Then follow spoken greetings in a wide variety of languages, and a so-called 'sound essay' consisting of sound effects of water, volcanoes, living beings, and technology such as a rocket launch. The record concludes with Earth's greatest hits, a musical selection ranging from tribal songs to Bach, Beethoven and Chuck Berry. One wonders what the hypothetical extraterrestrials will make of it all.

How might we first learn that we are not alone? One day we might intercept a probe-borne 'culture capsule' from another civilization, perhaps in the form of the interstellar messenger probes suggested by Ronald Bracewell. We might find a commemorative plaque at the site of an alien landing on another planet, or the

Moon, similar to the plaque attached to the Apollo 11 lunar module. There are, of course, those who think contact has already been made with UFO's and that visits have been frequent; the hard evidence, however, is still lacking. Probably the first real evidence of life elsewhere will be in the form of radio waves, the easiest way of sending information over great distances.

Whoever we contact in space (assuming the message does not come from too distant a star) is almost certain to be more advanced than we are, so we stand to learn a great deal from the extraterrestrials. Secrets of new energy sources, medical breakthroughs, and even the key to peaceful coexistence are some of the suggested benefits from communication with advanced beings. They may also be able to tell us how and where to get in touch with more civilizations, so that we become plugged in to some kind of galactic telephone network.

It is hard to imagine the sensation that the first announcement of contact with extraterrestrials will cause. Even if the message is incomprehensible, or if it says nothing of interest at all, its very existence will have answered the question: Are we alone? And knowing that advanced civilizations exist in space will be encouraging news to a young civilization such as ourselves, embattled as we are with the problems of technological development. It will demonstrate that such civilizations can survive without destroying themselves. Perhaps we will be able to emulate them. Contact with extraterrestrial life could mean the end of our childhood.

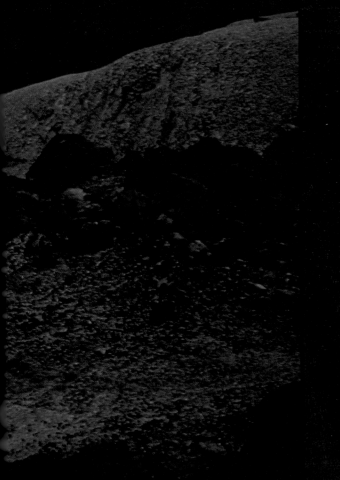

# 7/INTO THE DEPTHS

Humanity has long been awed by its seeming insignificance in the framework of the Universe. But just how *insignificant* we really are emerged only in the 1920's, when astronomers began to realize for the first time that all the stars we see at night – and millions more – are in our cosmic back yard, members of our own galaxy, the Milky Way (seen in the reconstruction on the left as if from a planet at the edge of a distant globular cluster). Beyond our Galaxy, 20 times its own diameter away, lies another galaxy, 300 billion stars wheeling like a vast Catherine-wheel in space. This galaxy, too, is just a next-door neighbour compared with the distant communities of galaxies that stretch out thousands of millions of light years in all directions. Moreover, the whole vast system of galaxies is expanding. But, will the expansion last for ever?

Complete acceptance of the fact that some astronomical objects are outside the Milky Way system came only in the mid-1920s. During the previous seventy years it had been established that many nebulae have a spiral form and the spectra of such objects suggested that they are collections of stars, but the distance to them was unknown. Novae were recognized and their apparent brightness suggested great distances, but just how great was uncertain, since the distinction between novae and supernovae was not yet clear. Only with the advent in 1918 of the 2·5 m (100-inch) Hooker reflector on Mount Wilson was it possible to detect individual stars in the nebulae and for approximate distance measurements to become practical. The conclusive step came in 1924 when Edwin Hubble used observations of Cepheid variables in several nearby nebulae to determine accurate distances. It was coming to be realized that the distribution of nebulae in the sky, in which they seemed to avoid the plane of the Milky Way, is an observational effect caused by interstellar obscuration and is not due to an uneven distribution in space, which would have required them to be physically associated with the Milky Way system. These results were related to the earlier concept of *island universes*, isolated separate star systems which later came to be called galaxies.

The word nebulae is now generally restricted to interstellar clouds of gas and dust. The term island universes is not now used, the *universe* is defined as the totality of observable things.

It is now customary to write Galaxy, with a capital G, when referring to the Milky Way system and to use small g for other galaxies. Generally, although not quite always, use of the adjective **galactic** is restricted to our own Galaxy, with increasing use of the new word **galaxian** for another galaxy or galaxies in general. Following on from this, **extragalactic** means anything outside the Galaxy, including other galaxies, and **intergalactic** is used to refer to what is between the galaxies.

In the past thirty years or so, radioastronomy has become just as important as optical astronomy in extragalactic studies. Indeed the two are largely complementary.

## Hubble's classification of galaxies

The first classification of galactic types was made by Edwin P. Hubble in the 1920s and is still in use, although it has been modified slightly. It is morphological, that is, based on the appearance of galaxies, and contains three major classes: galaxies with no evident internal structure, which appear elliptical (**elliptical galaxies**); those with a thin, flat disc exhibiting spiral arms and a condensed nucleus (**spiral galaxies**); and those which are neither of these (**irregular galaxies**). These classes have subdivisions and there is also a sequence of **barred spirals**.

Elliptical galaxies are classified by the shape of the ellipse seen in the sky. Mathematically, if we take $a$ and $b$ as the axes of an ellipse, the ellipticity, or flattening, is given by $(a-b)/a$. This runs from zero for a circle towards 1 for a very flat system. The galaxies are classified by multiplying this number by ten and then taking the nearest whole number. The number is preceded by the letter E. Classes observed run from E0 to E7 only, systems flatter than E7 no longer being ellipses but displaying a central condensation and a disc instead. Next come the lens-shaped galaxies S0 and SB0, which have discs but no spiral arms. They lie in intermediate positions in the continuous Hubble sequence from E to S and SB types.

The classification Sa, Sb, Sc for spiral galaxies depends on the size of the central condensation and the tightness with which the spiral arms are wound; they vary together. Sa type galaxies have arms that are tightly wound and large central condensations. Sb have more open arms and a smaller central condensation; Sc have very open arms and a very small nucleus. The amounts of interstellar gas and dust increase in sequence from Sa to Sc. For barred spirals, SBa and so on, the structures are the same except that the arms emerge from the ends of a prominent bar passing through the nucleus rather than the nucleus itself. Unlike an elliptical, the classification of a spiral galaxy does not depend on its orientation in the sky.

Two irregular galaxy types are recognized. The more common type Irr I, which includes both the Magellanic Clouds, clearly follows in sequence after Sc and SBc, with the break-up of arms into a confused structure and the presence of still more dust and gas. The second type, Irr II, is rarer; they are very different from Irr I galaxies, being peculiar, chaotic objects, highly obscured and reddened by internal dust. They lie outside the general sequence of galaxies, and often display strong activity.

The Hubble sequence is based purely on appearance, although there are systematic physical variations. The relative number of bright blue stars and the content of gas and dust increase steadily from E and S0 galaxies through the sequence of S or SB to Irr I.

The sequence is continuous, but it is believed that galaxies do not evolve along it, in either direction. One reason for this is that galaxies of all classes contain highly evolved stars (red supergiants), so they must all be around the same age, at least $10^{10}$ years old. Secondly, giant elliptical galaxies are more than ten times as massive as the largest spiral galaxies, and it would not be easy for the difference in mass to be gained or lost. Further, the observed rotation rates could not easily be changed.

Galaxies to the left in the sequence (E and S0) are called **early** and those to the right are called **late**.

Since the galaxies described in the Hubble sequence are all relatively nearby, the times taken for their light to reach us on Earth are much shorter than their ages. There is no question, then, of their appearance being affected by changes occurring within the long ages of a cosmological time scale.

## The modified classification scheme

Over the years various astronomers have introduced more detail into Hubble's original classification.

In the sequence between E7 and Sa, the So class may be conveniently divided into three (and similarly for SBo). Gérard de Vaucouleurs uses the labels So⁻, So°, So⁺, and precedes these by a class E⁺. Considering the relative dimensions of disc and nucleus, the So sequence can be regarded as running parallel to the S sequence, but of course without any spiral arms and containing much less gas. In addition, Sidney van den Bergh has described a sequence of gas-poor **anaemic spirals**, intermediate between S and So, which are common in clusters of galaxies.

The original classes Sc and Irr I cover a wide range in appearance and each has been split into two, Sc and Sd, Sm and Im. Sm contains irregular galaxies with slight but definite traces of spiral structure: in this system, the Large Magellanic Cloud is SBm. Im contains irregular galaxies without such structure: the Small Magellanic Cloud is Im. Both clouds have distinct bar structures.

Combination symbols, such as Sab, are used for intermediate types; (s) may be added if the spiral arms start in the nucleus and (r) if they start in a ring around the nucleus, as Sbc(s), SBa(r), with (rs) as intermediate type. De Vaucouleurs calls the ordinary spirals SA, to match the notation SB for barred spirals, with SAB intermediate between them. There is, thus, a continuous range covering three aspects, o-a-b-c-d-m, A-B, (r)-(s). Further, spiral arms of similar structure can differ in strength; some galaxies have thick, massive arms while in others, of the same Hubble type, they are thin and filamentary. They are sometimes distinguished by a subscript m or f.

The astronomer William Morgan recognized a special class of giant ellipticals, in 1958, which he called cD galaxies. These are very large, bright ellipticals with extended outer envelopes. Fre-

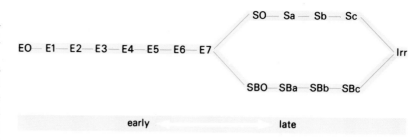

The Hubble diagram of galaxy types.

quently the largest galaxy in a rich cluster is a class cD; also most are strong radio sources. A couple of years later, it was found that the appearance of an Sb, Sc or Irr I galaxy is related to its luminosity. The most luminous galaxies have the longest and most fully developed spiral arms. The luminosity classes I to V, in decreasing order of brightness, were introduced. However, for Sb galaxies only classes I, II and III are used, for it seems that all the intrinsically faint spiral galaxies are of class Sc. Also, there are no class I irregular galaxies.

Luminosity classification gives a simple method for obtaining the relative distances of large numbers of spiral galaxies.

Spirals amount to about 75 per cent of the brightest galaxies, ellipticals and So 20 per cent and irregular 5 per cent. The relative numbers vary with limiting magnitude, however, for there are very many dwarf elliptical galaxies (sometimes counted as a distinct class dE) and also more irregular galaxies of low luminosity. For galaxies as a whole, therefore, the number of ellipticals probably exceeds 60 per cent and numbers of spirals and irregulars are approximately in the ratio 3 : 1.

The three groups, the ordinary spirals S (or SA), the barred spirals SB, and the intermediate group SAB, are present in about equal numbers.

Clearly, Hubble's classification scheme has undergone a major transformation as a result of more refined observations. The new generation of telescopes, including the future space telescope, may well lead to further changes.

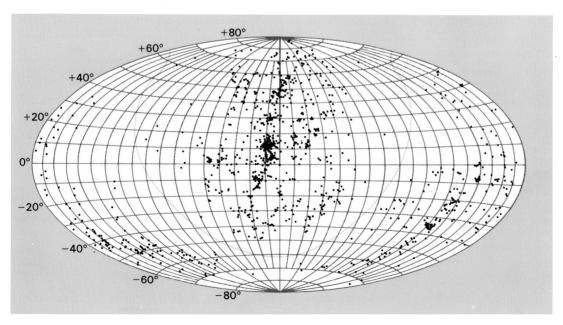

Positions in the sky of the thousand brightest galaxies, plotted using an equal-area projection of the entire sky using right ascension and declination as coordinates. The curved line represents the galactic equator. The distribution of galaxies avoids directions near the plane of the Galaxy, and also shows a strong tendency towards clustering.

Four typical galaxies. *Above*: the elliptical galaxy NGC 205, Class E5, a companion to the Andromeda galaxy. *Left*: the Sb galaxy NGC 4569. *Below left*: the Sc galaxy NGC 4565, which we see edge on. Notice the strong obscuration by dust, in a narrow belt. *Below right*: the barred spiral SBb, NGC 1300.

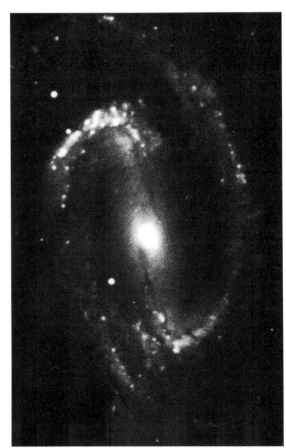

## The redshift

It is found that for all galaxies, apart from a few of the very nearest ones, the spectral lines are shifted to longer wavelengths. For optical lines, this means a shift towards the red. The more distant galaxies have larger redshifts, and the exact relationship between redshift ($z$) and distance – **Hubble's law** – was established by Hubble in 1929. It states that the redshift of an extragalactic object is equal to its distance multiplied by a constant. Mathematically it can be written in the form

$$cz = H_0 D$$

where $c$ is the velocity of light and $d$ is the distance corresponding to redshift $z$. The constant $H_0$, is now called Hubble's constant. (If $\lambda_0$ is the wavelength at the source and $\lambda$ the wavelength observed, $z = (\lambda - \lambda_0)/\lambda_0$. The value of $z$ is the same for all lines in the spectrum.) If the redshift is a Doppler shift caused by motion away from us, it is evidence for the expansion of the universe.

The rate of expansion may have changed as the universe has evolved; $H_0$ is the value at the present epoch, corresponding to all but a very few of the most distant observable galaxies. Velocities are measured in kilometres per second and distances in megaparsecs (Mpc), so it is convenient to express $H_0$ in the units km per second per Mpc. Hubble's own best estimate of $H_0$, given in 1935, was about 530, but we now accept that $H_0$ is much smaller than this, because, since 1935, estimates of the distances $d$ of galaxies have increased. This revision of the distance scale has happened in many small steps over the intervening years, although the largest single change came in 1952 when Walter Baade realized that the Cepheid and RR Lyrae variable stars do not have the same period–luminosity relation. As a result of Baade's revision, the extragalactic distance scale was doubled, and, whereas before 1952 the Andromeda galaxy (M31) was thought to be distinctly smaller than our Galaxy, since 1952 it has been known to be distinctly larger.

The use of the most modern instrumentation enables us to observe even more distant galaxies and obtain their spectra. At the time of writing the most distant known galaxy (as distinct from a quasar) is one associated with the radio source 3C 324, where $z$ is 1·21, but the vast majority of galaxies which can be studied easily have redshifts lower than 0·1.

Hubble's law relating distance to the observed redshift means that the scale of galaxian distances can be expressed by stating the numerical value of Hubble's constant $H_0$. It is appropriate to mention here that astronomers use a unit of distance called the *parsec* (pc). This is a distance of about 3·26 light-years. For very large distances, say those between galaxies, the parsec is used in preference to the light-year. One million parsecs is called a megaparsec and is written as 1 Mpc.

A recent systematic study to determine $H_0$ has been made by Allan Sandage and Gustav Tammann. According to them, $H_0$ is 55 km per second per Mpc, with a probable error of about 10 per cent. This is the best value available at present, in the sense of having the smallest probable error, and it is quite widely accepted.

In 1973, Vera Rubin, Kent Ford and Judith Rubin, using a sample of distant Sc I galaxies, presented evidence for a variation of $H_0$ with direction in the sky, but their interpretation was questioned. More recently, Vera Rubin and Ford, with other colleagues, have studied a larger sample of galaxies and confirmed their earlier result.

Different values of $H_0$ in different directions could be produced by different rates of expansion of the universe in different directions, by large random velocities of groups and clusters of galaxies, or by a large random velocity of the Galaxy and the Local Group. In the last case, the effect of our motion would be that the apparent velocity of recession of galaxies towards which we are moving is less, producing the appearance of a smaller $H_0$ there than in other directions. Rubin and her colleagues favour this third interpretation, and conclude that the Sun is moving at $600 \pm 125$ km per second relative to the distant galaxies. After allowing for the Sun's motion around the centre of the Galaxy and the motion of the Galaxy itself, the velocity of the Local Group is found to be 454 km per second.

There is also evidence for systematic differences in the motions of relatively nearby galaxies seen in different directions, an effect which can be interpreted as due to systematic motions within what is called the Local or Virgo Supercluster. Recent observations suggest that the Local Supercluster and the Hydra-Centaurus Supercluster are being drawn towards an even more massive collection of galaxies, whimsically called the "Great Attractor".

## The Local Group of galaxies

Our galaxy, the Milky Way, is a member of the Local Group, which has nearly thirty members. Most of their distances are accurately known. The Group contains a mixture of galaxy types, but there are no conspicuous giant ellipticals or barred spirals.

It is hard to establish the precise Hubble class for our Galaxy because we are inside it and so do not easily see its large-scale structure. The Galaxy has several small satellite companions, the Large and Small Magellanic Clouds being the most prominent. The Andromeda galaxy (M31) also has several companions.

Radio observations show that the SMC consists of two distinct parts, with different radial velocities, separated by about 20,000 light-years. The nearer has been called the Small Magellanic Cloud Remnant (SMCR) and the farther the Mini-Magellanic Cloud (MMC). The SMC was possibly torn apart when it approached the LMC around 20 million years ago.

In 1968, two galaxies, called Maffei 1 and 2, were discovered by infrared observation. Maffei 1 is a large E, or So galaxy and may be a Local Group member; Maffei 2 is spiral and almost certainly outside the Local Group.

### Stars in galaxies

In 1944, soon after the introduction to astronomy of a red-sensitive photographic emulsion, Walter Baade noticed that the bright stars in the spiral arms of the Andromeda galaxy are blue, while those in the nucleus are red; with only blue-sensitive plates this had not been observable. This led to the idea of population types. Population I contains blue stars found in spiral arms. Population II contains evolved objects, in particular red supergiants found in elliptical galaxies, globular clusters and the nuclei of spiral galaxies.

Even in the nearest galaxies only the brightest individual stars can be studied, so the stellar content of a galaxy is investigated with the spectrum of the total light, **integrated starlight**. Galaxy colours change through the Hubble sequence, reflecting an increase in numbers of Population I stars.

### Main members of the Local Group

| Galaxy (constellation) | Type | Distance (pc) | Size (pc) |
|---|---|---|---|
| Our Galaxy | Sb | – | 30,700 |
| Large Magellanic Cloud | SBc | 49,000 | 7050 |
| Small Magellanic Cloud | I | 61,300 | 3070 |
| Draco | E | 67,500 | 310 |
| Ursa Minor | E | 67,500 | 310 |
| Sculptor | E | 85,900 | 705 |
| Fornax | E | 168,700 | 1650 |
| Leo I | E4 | 230,000 | 610 |
| Leo II | E1 | 230,000 | 310 |
| NGC 6822 | I | 490,800 | 2300 |
| Wolf–Lundmark | E5 | 490,800 | 1530 |
| NGC 205 | E5 | 644,000 | 2390 |
| NGC 221 | E2 | 659,500 | 705 |
| NGC 147 | E4 | 659,500 | 1410 |
| NGC 185 | E0 | 659,500 | 1000 |
| NGC 224 (Andromeda) | Sb | 674,800 | 49,100 |
| IC 1613 | I | 736,200 | 3070 |
| NGC 598 | Sc | 828,200 | 14,100 |
| LGS 3 | I | 828,200 | 490 |
| Maffei I | E | 1,012,200 | – |

### Nearby clusters of galaxies

| Cluster name | Distance (Mpc) | Diameter (Mpc) |
|---|---|---|
| Sculptor | 2·4 | 1·0 |
| Messier 81 | 2·5 | 1·8 |
| Canes I | 3·8 | 1·9 |
| Messier 101 | 4·6 | 1·8 |
| NGC 2841 | 6·0 | 1·6 |
| NGC 1023 | 6·3 | 2·2 |
| NGC 2997 | 7·6 | 1·9 |
| Messier 66 | 7·6 | 1·0 |
| Canes II | 8·0 | 3·0 |
| Messier 96 | 8·3 | 1·6 |

*Opposite*: The large Magellanic Cloud. Note the strong bar structure and the slight suggestion of weak spiral arms.

ellipticals | lenticulars | spirals | irregulars
(lens-shaped)

*Above*: A three-dimensional representation of the Vaucouleurs' classification scheme.

*Right*: A cross-section through the above figure, near the region of the Sb spirals, showing transition cases between the ordinary (SA) and barred (SB) spirals and between those with (r) and without (s) inner rings.

*Opposite page, bottom*: The range of distance over which various extragalactic distance indicators can be used. The step-by-step nature of distance determination is clearly seen.

## The structure of galaxies

The true three-dimensional shapes of elliptical galaxies are presumed to be oblate spheroids with flattening produced by rotation, although recent measurements of rotational velocities suggest that this could be an oversimplified picture. It is found that there are not many truly spherical galaxies. Also, there are no elliptical galaxies with true ellipticity flatter than 0·7: flatter galaxies have discs and nuclei.

Studies of the distribution of luminosity in elliptical galaxies show an increase of brightness towards the centre, indicating an increase in the density of stars there. Moreover, as the intensity falls off faster near the outer boundary, the size of ellipticals can be determined relatively accurately.

The basic structure of a spiral galaxy is a nucleus plus a disc containing spiral arms. The spiral arms contain Population I stars, while Population II stars are found in the nucleus and distributed evenly over the disc. The relative size of the nucleus and the degree of openness of the arms are, as we have seen, classification parameters. Within the

Recently, a few nearby elliptical galaxies have been found to contain small populations of blue giant stars. These observations disturb the conventional idea that ellipticals contain only old material.

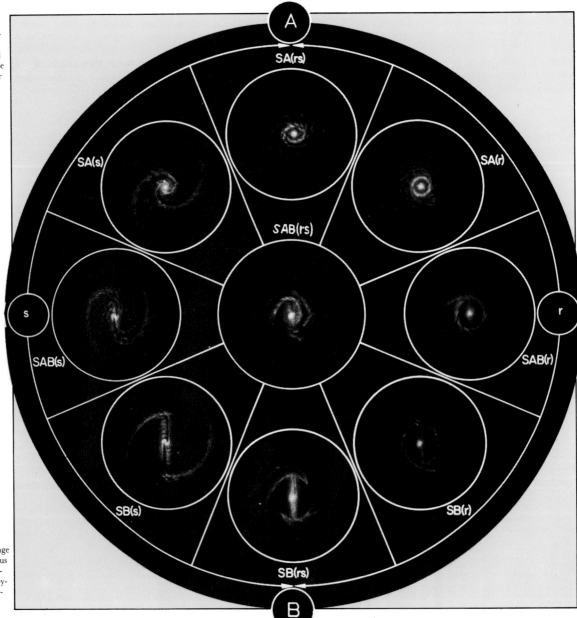

nucleus there is a distribution of luminosity similar to that in an elliptical galaxy. The disc, averaged round in angle to smooth out the spiral structure, is bright beside the nucleus and becomes fainter quite rapidly with radius, but there is no sharp boundary. A longer photographic exposure reveals material farther out, while radio observations reveal material at twice the radius detectable optically. It is not easy, therefore, to determine the full dimensions of spiral galaxies.

Galaxy diameters range from about 3000 light-years or less for dwarf ellipticals up to more than 150,000 light-years for the brightest cD galaxies seen in rich clusters. The largest galaxies have a luminosity of about $10^{11}$ Suns. Recent observations have revealed even larger galaxies, including one, Malin 1, an isolated spiral that is probably about 750,000 light-years in diameter and may be very young.

### Interacting galaxies

Systems which have a large galaxy with several small companions, as with our Galaxy and M31, are common. In addition, there are other cases where two or more galaxies of nearly equal mass are closely associated. Galaxies of different classes are found together, supporting the idea that they have similar ages.

When the galaxies are close enough, optical filaments, bridges and other structures are observed optically and at radio wavelengths. Computer models, based on gravitational interaction, have been constructed which indicate that many of the strange shapes observed are produced by mutual tidal distortion of the two galaxies. It is possible, though, that some of the other distorted shapes could be due to activity in the nucleus of the galaxy.

One explanation of the warping commonly found in the discs of spiral galaxies is that it is a tidal distortion. For our Galaxy, the observed warping could have been produced by the Large Magellanic Cloud. However, for this to have happened it would need to have passed close to the Galaxy some $5 \times 10^8$ years ago and also to have a larger mass than is otherwise supposed. A problem using the hypothesis of tidal interaction is that warping is observed in some galaxies which have no known companions. One way round this is to assume that unseen intergalactic material is responsible for the warping.

### Clusters of galaxies

The distribution of galaxies is not random. There is a strong tendency towards clustering. A cluster is a real physical clumping of galaxies, with a greater number of galaxies inside a cluster than outside. The diameter of the cluster may be as large as 10 Mpc. Clusters range from groups with ten or twenty known members, similar to the Local Group, up to rich clusters containing several thousand galaxies. Tens of thousands of clusters can be identified in sky survey photographs; the most conspicuous ones are named after the constellations in which they lie.

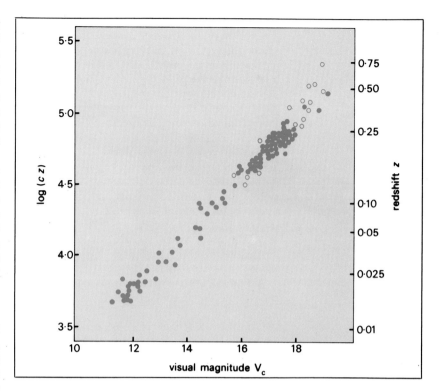

Clusters are classified as **regular** or **irregular**. All regular clusters are large, with thousands of members, and have a spherical shape with a concentration of galaxies towards the centre, where there is often a cD galaxy; a well-studied example is the Coma cluster.

Irregular clusters extend in a continuous range from small groups to large, rich clusters of galaxies. Unlike regular clusters they have little symmetry and little concentration towards a centre. Examples of irregular clusters are the Local Group and the Virgo cluster, which is the nearest large one, which has about 2500 known members. The centre of the Virgo cluster is about 20 Mpc distant from the Sun.

*Above*: The Hubble relation for the brightest galaxies in large clusters. z is the redshift and Vc is a corrected apparent visual magnitude. If it is assumed that all these galaxies have the same absolute magnitude, Vc is related to distance. The open circles come from more recent observations and suggest that for larger z the plot departs slightly from a straight line.

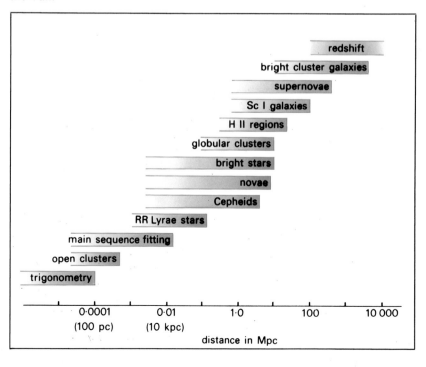

Compared with the statistics for galaxies in general, irregular clusters contain slightly fewer spiral galaxies, and regular clusters very few, particularly in the central regions. Some years ago, it seemed that the probable explanation was that in the denser clusters direct-hit collisions between galaxies would be more frequent, with the result that gas and dust would be swept from the colliding galaxies, causing spirals to become ellipticals. However, recent research shows that such collisions would be too infrequent. It now seems probable that the gas and dust are pushed out of a galaxy as it passes through relatively dense intergalactic material in the central region of a cluster. There is, however, another possibility – that, for some unknown reason, fewer spirals formed in these higher density regions in the first place.

About half of the nearby regular clusters and a quarter of the irregular clusters contain radio galaxies; indeed, about 20 per cent of all radio galaxies lie inside rich clusters. It may be that many of the more distant radio galaxies, which appear separate, belong to clusters in which the other galaxies are too dim to be seen. In any case, the figures indicate that giant elliptical galaxies have about the same chance of being strong radio sources whether they are in rich clusters or not.

Roughly a third of the nearby regular clusters have been identified as X-ray sources. They include most of the optically identified, extragalactic X-ray sources. Generally, the X-rays are emitted from an extended region in the cluster centre, larger than a galaxy, and in some cases there is also emission from a central active galaxy. The most likely source of X-ray emission is a very hot ionized gas (or plasma) lying in between the galaxies of the cluster, an intergalactic and intracluster gas, with a temperature of about $10^8$ K and density about 100 electrons per cubic metre.

Clusters do not appear to be distributed at random, and this is generally interpreted as good evidence for clusters of clusters, or superclusters. The Local Group is believed to be part of a supercluster centered on the Virgo cluster. This is known as the **Local Supercluster** and is about 240 million light-years across. Going further, there are even indications of an enormous supercluster complex even larger than the "Great Attractor", but such vast groups pose problems for theories of galaxy formation.

### The mean density of the universe

The rate of expansion of the universe is becoming slower because of gravitational forces. If its average density is larger than a certain critical value, gravity will take over and expansion will eventually stop, to be followed by contraction; the universe is said to be **closed**. If the density is less than this critical value, expansion will continue for ever; the universe is **open**. Using the value of 55 km per second per Mpc for Hubble's constant, this critical density is equivalent to 3–4 hydrogen atoms per cubic metre. A reasonable estimate for the matter present in galaxies gives, when averaged through space, only one-fiftieth of this value. Supposing, though, that the mass discrepancy in clusters were caused entirely by underestimating the masses of galaxies (which is possible but unlikely), then the averaged mass would be brought up to about one-third of the critical value. It seems reasonably certain that galaxies and whatever it is that binds clusters together would not be sufficient to close the universe. Some other material, in between the clusters, is required.

It is possible that gas exists in the space between clusters. Such gas could be electrically neutral or ionized, and, because of the expansion of the universe, it should display a redshift. Observations show that if there is neutral gas spread throughout space, it can be there only in tiny amounts, and that even if it were concentrated into dense clouds, there could be no more than about one-third of the critical density. One instance of a fairly massive cloud of neutral hydrogen is known at a distance of about 10 Mpc. The indications are that this is at least 100 kpc across, and although the average density is low it appears to contain about as much mass as a galaxy. Whether other examples exist remains unknown at present.

We do observe an X-ray background coming apparently from every direction over the whole sky, and this can reasonably be interpreted as emission by an intergalactic ionized gas at about $10^8$ K. There is an alternative explanation for this radiation, which is that it comes from a host of faint discrete sources. If this were so it should be possible to detect fluctuations in the amount of radiation over small areas of sky. At present observations are not precise enough for us to establish whether such fluctuations exist. In any case, it seems likely that the density of intergalactic gas required to produce

The relation between redshift, distance and light travel time, using a very simple theory. A value 55 has been adopted for Hubble's constant $H_0$.

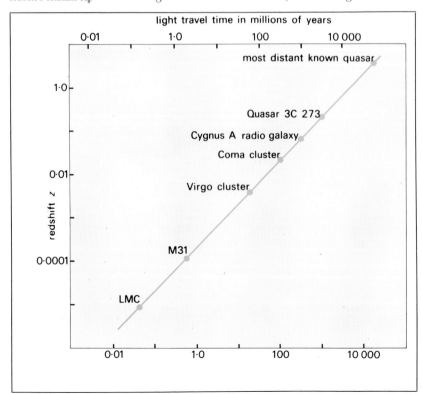

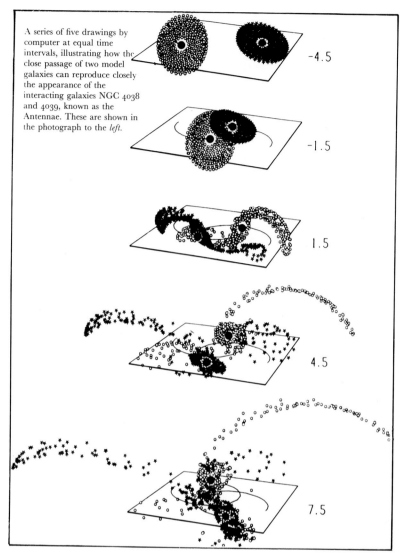

A series of five drawings by computer at equal time intervals, illustrating how the close passage of two model galaxies can reproduce closely the appearance of the interacting galaxies NGC 4038 and 4039, known as the Antennae. These are shown in the photograph to the *left*.

-4.5

-1.5

1.5

4.5

7.5

the observed radiation is less than the critical density, although it is just possible that it could equal it. Moreover, there are several problems concerned with maintaining the high temperature required for the radiation and with the effects of the hot gas on clusters and galaxies embedded in it, which make it appear necessary that the density of a $10^8$ K intergalactic gas must be significantly below the critical value.

Another possibility is the presence of intergalactic dust. This would redden the light from galaxies, but no such reddening is observed. The amount of any dust present must be far too small to contribute significantly to the total density. Indeed, this is to be expected, because intergalactic material probably contains scarcely any atoms heavier than hydrogen and helium, since heavy atoms are created by nuclear processes inside stars. Moreover, this expected lack of heavy elements also rules out the presence of very large numbers of larger solid objects, of sizes between a few centimetres and planetary sizes, even though

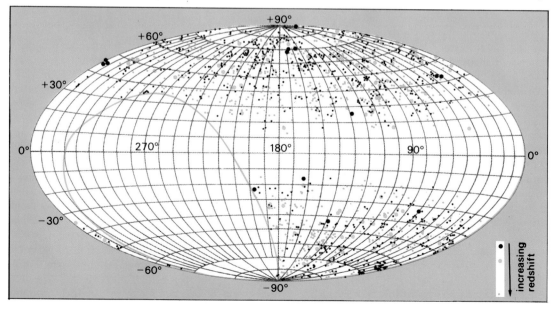

increasing redshift

The distribution of clusters of galaxies in the sky, plotted in galactic co-ordinates with the galactic anticentre near the centre of the plot. The large empty oval was not covered by this survey. Clusters show a clear tendency themselves to cluster together. Notice also the zone of avoidance on either side of the galactic plane.

they are not ruled out from an observational point of view.

Intergalactic populations of subluminous objects – very cool main sequence stars, white dwarfs, neutron stars and black holes – which could in total contribute more than the critical density are not ruled out by observation. However, it appears unlikely that a large population of low mass main sequence stars could exist outside galaxies, while the other objects form as end products of stellar evolution, so they are also to be expected in substantial numbers only inside galaxies. A final possibility of meeting the mass deficiency is that there might be significant quantities of neutrinos or **gravitons** (gravitational wave packets), but there is no possibility of detecting them if they are there.

Because of all the many uncertainties, the question of whether the mean density is greater or less than the critical value is still unresolved. Nevertheless, since all forms of matter which can be studied with accuracy contribute substantially less than the critical density, and most other forms of material appear implausible, it is fairly widely felt that the total density is probably below the critical value, in which case the universe will expand for ever.

### Galaxy formation and evolution

We now believe that the age of the universe is around $1 \cdot 5 \times 10^{10}$ years. Observations of distant quasars show that they contain elements other than the hydrogen and helium formed in the Big Bang. They must therefore have contained at least one earlier generation of stars. The indications are that galaxies (or quasars) existed when the universe was only about 15 per cent of its current age. Explaining galaxy formation, however, remains one of the major problems in theoretical cosmology and there are a number of conflicting theories.

The basic difficulty comes in deciding whether the first objects to form were galaxies or clusters of galaxies. Either solution encounters problems. If it is assumed that galaxies form first, and that they then come together in both clusters and super-clusters, it is very difficult to account for the predominance of elliptical galaxies in very dense clusters, because this would imply that galaxies "knew" what type of cluster they would be joining.

It seems perhaps simpler to assume that following the Big Bang, very large-scale aggregations of matter formed. These sheets, "pancakes" and filaments then broke up into individual galaxies. Unfortunately, there are now theoretical problems with the time-scale involved. If the largest structures formed first, then most galaxies should still have been forming when the universe was only half its present age. It appears, however, that the majority of galaxies must have been formed well before that time, even though there are indications that galaxy-formation may still be occurring now, albeit at a very low rate.

Some of the latest developments in cosmological theory give support to the second general picture. Theoretical structures known as "cosmic strings" may form "loops" – which may be considered to be "flaws" in space–time that appear to cluster hierarchically. Before they disappear, these loops exert strong gravitational forces, which may act to promote the formation of clusters of galaxies.

Accepting the idea that the galaxies all formed from the same material at about the same time, the differences between them must have arisen from different evolutionary developments after they formed. Two major properties of a galaxy are its mass and its rotation, but since all the Hubble types have a wide range of mass, it seems that mass is not decisive. On the other hand, a galaxy's rotation appears to be linked with Hubble type and can, in outline at least, explain the different developments. The flattening of an elliptical galaxy can be related to its rotation, while in a disc galaxy, faster rotation means a slower collapse, so that up to the present time less of the interstellar gas has been used forming stars.

However, some consider that the difference between elliptical and disc galaxies requires further explanation. It could be that outside influences have in some way affected the manner in which star formation proceeded, or possibly it has had something to do with activity in the nucleus. Type So galaxies may be spirals which have been swept clear of gas, either through the pressure of the intergalactic gas in clusters or, less frequently, by direct collision between galaxies.

After forming, a rotating gas disc becomes thinner with time. Gas passing through the plane of the disc collides with other gas and so tends to lose its motion perpendicular to the plane. Stars, however, can pass freely through the disc, and their distribution does not thin out. This scenario is consistent with the distribution in our Galaxy of stars of different population types where the younger populations (stars which formed later) have thinner disc distributions.

General theories of the formation and evolution of galaxies have a long way to go before they can be related to real galaxies. It is not clear, for instance, whether all galaxies go through a phase of being active, perhaps early in their lives, although this is a question of fundamental importance.

### Active galaxies

In a normal galaxy, everything seems to be stable and in equilibrium. An active galaxy, in contrast, is one where there is strong emission, and rapid variations of intensity. Active galaxies have been found in two different ways, by radio and optical techniques. These two groups overlap but do not coincide. Strong radio sources may be optically normal or optically peculiar, and some peculiar galaxies are not strong radio sources.

In 1943, Carl Seyfert published a list of six unusual galaxies, with a small, bright nucleus and faint arms. About 100 galaxies are now classified as **Seyfert galaxies**. The optical spectrum of the nucleus contains emission lines not seen at all strongly in normal galaxies, but commonly observed in gaseous galactic nebulae. The spectral lines are particularly broad, which implies expansion out from the nucleus at high velocities, around 500 km per second.

Two classes of Seyferts have been identified. In Class 1, the hydrogen lines are broader than the forbidden lines, so emission is from different regions; in Class 2 the lines have the same width. Class 1 emission lines are weak compared with the continuum emission from the nucleus, and in Class 2 they are strong. The nucleus appears smaller in Class 1 Seyferts. In many ways the nuclei of Class 1 Seyferts resemble quasars. Most Seyfert galaxies radiate strongly both in infrared and ultraviolet light. Although all emit radio waves more strongly than ordinary galaxies, only a few are very strong radio sources. Several Class 1 Seyferts are strong X-ray sources.

Almost all Seyferts appear to be S or SB galaxies with active nuclei, and it is estimated that about 1 per cent of all spiral galaxies are Seyferts.

N galaxies are basically the same as Class 1 Seyferts, but the nuclei are brighter and more compact and the surrounding galaxy is less clear. Both have larger redshifts than ordinary galaxies. N galaxies were identified as a separate optical class by William Morgan in 1958, and came to be studied particularly in the 1960s when optical surveys were made of radio sources.

An extensive survey of galaxies which are bright in the ultraviolet has been carried out since 1967 by B. E. Markarian at the Byurakan Observatory in Soviet Armenia. His lists so far contain about 700 objects, of which about 10 per cent are Seyfert galaxies; indeed, it is through his work that most known Seyferts have been discovered.

Markarian galaxies are of two major types, one where the ultraviolet emission is from a bright nucleus, and the other where it comes from the whole galaxy. The galaxies with bright nuclei include the Seyferts but most are of a new type. They appear to have an excess of hot stars in the nucleus, which causes the emission of narrow spectral lines. There is also a variety of other objects, ranging in size down to small dwarf galaxies which, unlike dwarf ellipticals, have strong ultraviolet radiation, and so must contain very young, hot stars and ionized gas.

Starting in the 1930s, Fritz Zwicky prepared lists of objects appearing in the first Palomar Sky Survey photographs. They are only just distinguishable from stars, and Zwicky called them **compact galaxies**. They include objects now known to be at very different distances. A few of them are probably nearby dwarf galaxies. Many, however, appear to be ordinary galaxies seen very far away, while some are N- and Seyfert-type objects.

## Quasars

Quasars or QSOs (quasistellar objects) are defined as starlike objects whose optical spectra contain bright emission lines with large redshift, although marginal cases, such as 3C 48 which has weak nebulosity or 3C 273 with a jet feature, are not excluded. Both of these objects are now known to be surrounded by considerable nebulosity in the form of an associated galaxy. Usually the optical light varies irregularly over a time scale of months, there is strong ultraviolet radiation, the emission lines are broad, and often there are also narrower absorption lines. Many quasars are strong radio sources, of very small angular size. A radio-strong quasar is denoted by QSS (quasistellar-source)

Although the first identification of a radio source with a starlike object was in 1960, it was three years before the emission lines could be identified. In 1963, Maarten Schmidt recognized them as familiar lines but with unprecedentedly large redshifts. Within the next two years Allan Sandage recognized the existence of radio-quiet quasars. It has been estimated that around a million QSOs are detectable and that the QSSs are outnumbered about 100 to 1 by the radio-quiet objects.

Redshifts have been determined for hundreds of quasars. The distribution in redshift is smooth, and values range from $z = 0.04$ up to $4.43$, although most are below $2.4$. If the redshifts are related to the quasar distances by Hubble's law, then most quasars are very much more distant than ordinary

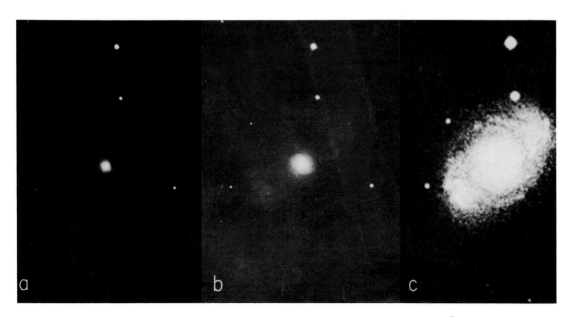

The Seyfert galaxy NGC 4151 in three photographs on the same scale but with different exposures. In (a) only the starlike nucleus is seen (making it look like a quasar); outer spiral structure is seen faintly in (b) and strongly in (c).

galaxies or even active galaxies. Most astronomers now accept that there is a continuous progression through Class 1 Seyferts, N galaxies and quasars and that their distances are indeed great. From Seyferts to quasars we progress towards relatively more intense nuclei. A quasar appears starlike because its distance makes the rest of the galaxy undetectably faint.

**BL Lac objects**

The object BL Lacertae, once classified as a variable star, was identified with a radio source in 1968, and now about 40 other objects like it are known. They are essentially similar to quasars and the nuclei of Seyfert and N galaxies, except that the optical spectrum of the compact object is a continuum; no emission lines are seen. Some appear in the nuclei of elliptical galaxies, some are associated with nebulosity, but most appear stellar on the best available photographic plates.

They are like more extreme versions of quasars and N galaxies. Almost all of those known are radio sources, but none is associated with a large extended source as many quasars are. None is yet identified as an X-ray source. They radiate most strongly in the infrared.

It has been suggested that the BL Lac objects are young quasars which have not yet ejected much plasma or gas. This would explain the lack of association with extended radio sources and also the absence of emission lines, believed in quasars to come from ejected gas outside the central region. On the other hand, the fact that some BL Lac objects are seen in elliptical galaxies has led to the suggestion that they are old active objects.

**Extragalactic radio sources**

The first radioastronomy surveys of the sky, made at various wavelengths in the late 1940s and 1950s, showed hundreds of discrete sources, but it required optical identification to establish how many are extragalactic, for from radio observation alone we have no certain knowledge of the distance. It turns out, however, that fewer than 10 per cent of the identified sources are galactic; these are mostly clouds of ionized hydrogen and supernova remnants, and are concentrated towards the plane of the Milky Way. Identified extragalactic sources are more evenly distributed over the sky and, although nearly a third of the brightest sources are not identified, it is clear from their distribution that almost all the unidentified sources are extragalactic too. Some very strong sources are associated with faint galaxies, and similar objects farther away will still be reasonably strong radio sources but invisible optically.

Extragalactic optical identifications which have been made are, in rough order of increasing absolute radio luminosity:

a. normal spiral galaxies. These are relatively nearby and have thermal radio emission similar to galactic sources. They are not classed as 'radio galaxies';
b. Seyfert galaxies;
c. certain bright early type (E and SO) galaxies. Most such galaxies are not detected as radio sources at all, but some of the largest ones, particular cD galaxies, are very strong sources. There is generally nothing special about the optical appearance. Often, the brightest galaxy in a cluster is a strong radio source;
d. N-type galaxies;
e. radio quasars, QSSs, if their redshifts are interpreted as indicating their distances.

**Structures of radio sources**

Some sources are compact, with a single component centred on the nucleus of the optical object, but most have an extended structure, typically in two similar lobes roughly symmetrical about the optical nucleus. Both structures are found to be associated with all the different types of optical object, although most compact radio sources are found to be associated with quasars or galaxies with bright nuclei. It should be noted that a particular radio source can be identified as a quasar only if it is related to a starlike optical object.

Some two-lobed sources are very large, extending over several Mpc. The normal structure within each component is a strong head with a tail extending back towards the centre, and quite often there are inner maxima in the intensity, much closer in than the outer lobes but aligned along the same axis.

In most cases, the radio waves show polarization, in amounts up to 20 per cent. Polarization is related to the strength and direction of a magnetic field, and in some sources the field is uniform over remarkably large volumes of space. The most powerful radio sources so far detected radiate about $10^7$ times more energy at radio wavelengths than the whole Galaxy does, while irregular variations within a few months are observed in many compact sources.

The relative linear sizes of a sample of typical extended radio sources. Sources of such different extent have essentially similar structures. Notice the small central component of 3C 236, with structure aligned with the outer lobes (*see opposite picture*).

DA 240

Cygnus A

3C 236

1 Mpc

× 300

central component

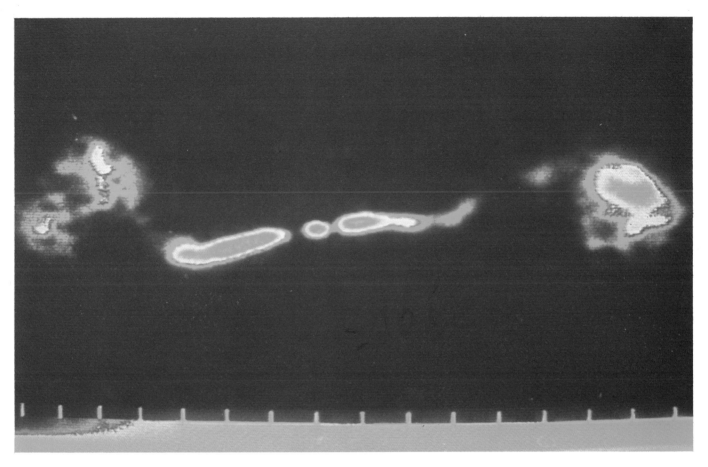

## The active galaxy M 87

M 87 is a giant Eo galaxy, about $10^{13}$ solar masses, and the third brightest member of the Virgo cluster. Optically it displays a strong jet, similar to the one seen in 3C 273. As the third brightest radio source in the sky, it is also known as Virgo A; it appears bright because it is very close to us, only about 15 Mpc away. At radio wavelengths it is seen to have two jets, one coincident with the optical jet and one opposite it, as well as a fainter outer halo. The core is very compact, emitting about 1 per cent of the radio energy from a region no more than about 0·1 pc across. M 87 is also an X-ray source.

Early in 1978, evidence from optical astronomy helped towards an understanding of the nucleus. It is found to be very bright in relation to the rest of the galaxy, compared with ordinary ellipticals, while the spectral lines reveal a sharp increase in the range of velocities in the nucleus. These optical observations, together with the observed radio structure, indicate not only an unusual concentration of mass at the galaxy centre, but also an energy source. The most plausible explanation is considered to be the presence of a black hole of about $5 \times 10^9$ solar masses. Although not at all conclusive, this is the strongest evidence available for the existence of a black hole in an extragalactic object.

## Energy sources for radio galaxies

The first energy source proposed for radio galaxies was the collision of two galaxies in a cluster, partly because the very strong source Cygnus A looks like two galaxies colliding. However, many radio sources appear optically normal, and in any case collision would not give either synchrotron radiation or a two-lobed structure. It is now believed that radio sources originate in violent events within single galaxies, which are related to the energy sources in quasars and active galactic nuclei.

Various possible sources for such violent releases of energy have been suggested. The energy release from the gravitational collapse of an object of some $10^6$–$10^8$ solar masses was one, but it would tend to happen too quickly and again not produce two lobes or the fast particles which give synchrotron radiation, and the collapsing body would probably be unstable anyway. Another was a multiple outburst of supernovae where, in conditions of high stellar density in a galactic nucleus, one supernova might trigger off another in a chain-reaction sequence; but again the energy release would be in all directions and too fast. A massive black hole in the galactic nucleus, as discussed above for M 87 and with less confidence for normal galaxies, has been proposed, but as the source of energy for an extended radio source, it presents similar problems. An opposite idea is that there could be a **white hole** in the nucleus, and that material for the entire galaxy could be emerging from it. This idea has received particular attention from Viktor Ambartsumian, who further suggests that entire clusters of galaxies could have emerged from such holes and still be dispersing.

Another class of theories, introduced by L. M. Ozernoy and developed by Philip Morrison and others, considers a rapidly rotating region in the

A 'photograph' made from radio observations of 3C 499, showing the extended lobes of radio emission on either side of the parent galaxy.

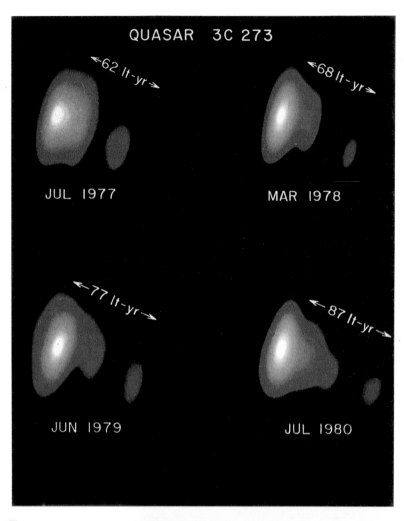

QUASAR 3C 273

←62 lt-yr→

JUL 1977

←68 lt-yr→

MAR 1978

←77 lt-yr→

JUN 1979

←87 lt-yr→

JUL 1980

*Above*: The apparent superluminal velocity of the quasar 3C 273. Over a period of 3 years it has appeared to expand by 25 light-years.

centre of the galaxy, of $10^8$–$10^{10}$ solar masses and called a **spinar**. Magnetic interactions at its outer edges would release energy. In a version worked out by Martin Rees, energy is released as a highly energetic plasma along the two ends of the rotation axis; thus a double structure is produced along a defined axis, as is observed. Such "jet processes" are coming to be accepted as important in many unusual objects, both in accounting for the double structure of many radio sources, and also on a smaller scale for individual objects within galaxies. (An example of the latter is the object known as SS 433, which can be explained by a compact object accreting matter from a companion in a binary system and giving rise to two jets of energetic particles which are ejected along its rotational axis.) With a spinar, plasma and energy are supplied in a continuous outflow from the nucleus. The origin of the hot plasma is another problem; again, it could involve a black hole.

**Apparent superluminal velocities of radio sources**

Some of the compact radio sources, are found to be double-lobed, similar to extended sources. The angular sizes of less than 0·01 second of arc correspond to actual dimensions of only a few parsecs. In at least four cases – three quasars and one galaxy – structural changes, which have taken place over several years, give an apparent velocity of expansion much greater than the speed of light. This is based upon the assumption that the distances are given by interpreting the redshifts in accordance with Hubble's law. As one of the

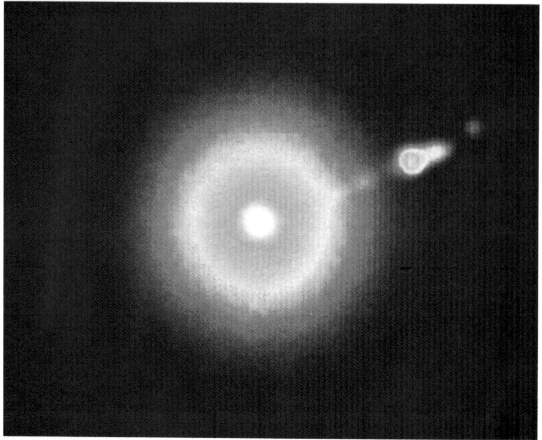

Computer-processed optical picture of the active galaxy M87, showing the various bright knots along its jet.

sources, 3C 120, is a galaxy, this cannot easily be questioned.

It is a fundamental result of modern physics that nothing can move faster than the velocity of light. This result is, indeed, an essential part of the theory of relativity. However an *appearance* can move faster than light, so long as no direct *physical* motion is involved. This is now thought to be the explanation of the phenomena observed, especially after examination in detail of one such source, the nearby quasar 3C 273. It is believed that jet processes are producing a beam of highly energetic electrons travelling at close to the speed of light. Under such circumstances, despite what might be thought at first, if the beam is directed towards us rather than across the line of sight, the material excited by the electrons can appear to move faster than the speed of light. In the case of 3C 273 the beam is probably within about 12° of our line of sight. The other instances almost certainly have the same cause, and statistical studies are likely to show that because we may expect the beams to be randomly orientated in space, only a few such examples will be visible to us.

### Does BL Lac = quasar = radio galaxy?

A new model, developed by the Dutch astronomer Peter Barthel, of the University of Groningen, strongly suggests that what appear to be three of the most energetic kinds of object in the universe – BL Lac objects, quasars and radio galaxies – are in fact the same. The observed differences may arise because we simply see them in different orientations. Both quasars and radio galaxies produce jets of high velocity electrons. Barthel's model suggests that we see a radio galaxy if we happen to observe the jets from the side, and a quasar is seen if we observe a jet almost end-on.

So, it looks, from the new model, as if radio galaxies are really under-cover quasars. When we look at a radio galaxy, what we may actually be observing is a quasar which is surrounded by a huge ring-like structure of dense dust clouds, with a diameter of a few hundred parsecs. The dust clouds can be considered to form the shape of a giant doughnut. Then it can be imagined that the jets of electrons emerge along the axis of the doughnut. If we happen to see the doughnut from the side, then the dust hides the bright core, and astronomers observe only the two jets stretching out to the distant radio lobes. This would lead to the object being classified as a radio galaxy. If the doughnut is turned so that we can look almost along the axis, then we can look to the central region and see the bright core – the quasar.

This model also helps to explain other observations that have long been puzzles. Quasars only ever seem to possess a single jet, despite the fact that they have two radio-emitting lobes which indicate that the core emits streams of high-energy particles in two opposing directions. The model accounts for this very well. In a quasar, astronomers are looking into the doughnut, so one of the jets is coming almost directly towards us, while the other is moving away from us. Relativity theory tells us that the radiation from fast moving objects is beamed, and thus seemingly enhanced, in the direction of travel, that is in a forward direction. So, the jet coming towards us should appear hundreds of times brighter than the receding one. It is also boosted by the Doppler effect. As a result, our current generation of telescopes reveal only the oncoming jet; they cannot detect the receding one.

In a radio galaxy, neither jet is directed toward us, so they should both be visible, but they will be much fainter than the single jet observed from quasars. It has been possible to detect both jets in some radio galaxies.

One thing that concerns astronomers is the amount of polarization found in the two outer lobes of quasars. The radio waves coming from both lobes should exhibit strong polarization. But it is generally the case that, in quasars, one lobe shows very little polarization. The best explanation for this is that the radiation from this lobe has travelled through a cloud of hot gas, which would tend to wipe out any evidence of polarization. This would be expected if we are looking at the quasar almost end-on. The radio waves from the lobe on what could be called our side of the quasar, would show strong polarization. The other lobe lies on the far side of the quasar, and hence on the far side of the hot gas in the doughnut surrounding the quasar.

The model also neatly ties in with observations of the even more powerful BL Lac objects. These appear as small, but very powerful, sources of light and radio waves. In fact they seem so powerful that the light all but obliterates the spectral lines from the object. The suggestion is that BL Lac objects are quasars seen almost exactly end-on. In other words, we could be looking straight along the axis of the jet, and its radiation is amplified by the relativistic beaming effect.

So it could be that what astronomers have for so many years thought to be three distinct types of object, are in fact all the same. It is just our vantage point in space that makes them appear to be so very different.

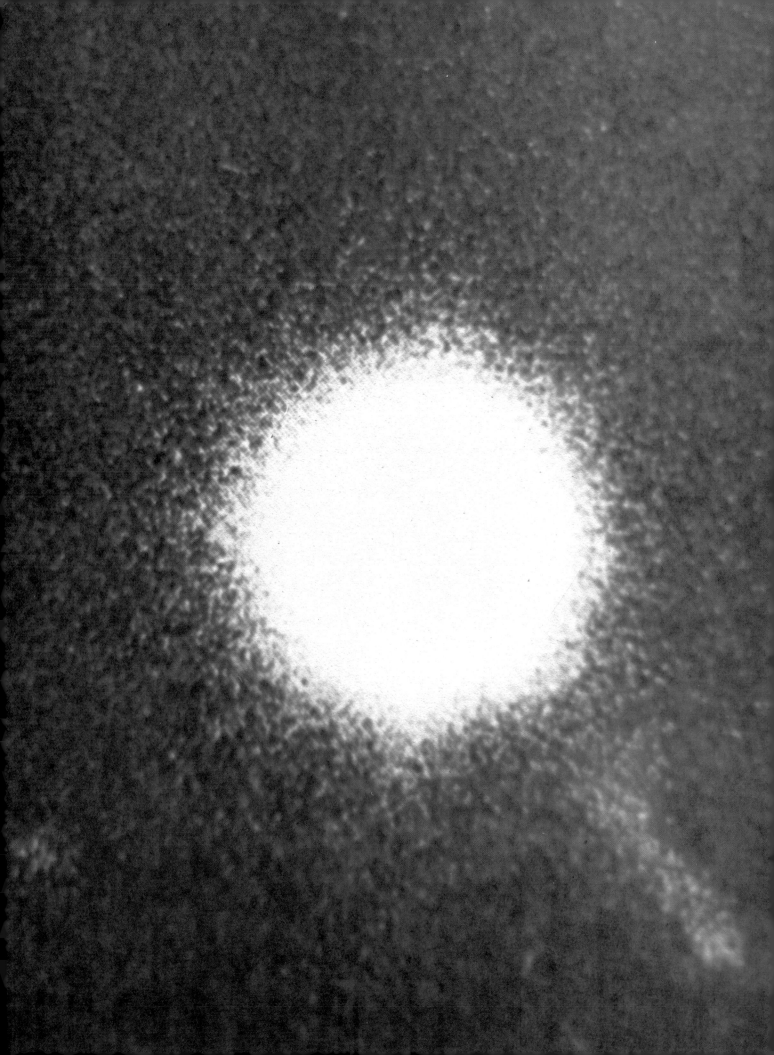

# 8/TO THE EDGE OF ETERNITY

By measuring the rate of expansion of the Universe, we can
now say that it came into existence no longer than 20,000
million years ago, in a monumental fireball that provided
the raw materials from which the galaxies, stars, planets,
and even we were formed. We can observe the effects of that
violent birth. The galaxies – as we know from studying
energetic, distant quasars like 3C 273 (left) – are fleeing
from us at speeds up to 80 per cent of the speed of light.
However, what was there before the Big Bang and what is
the fate of the Universe? Is it possible that the universal
expansion will be reversed by the pull of gravity, that all
matter will collapse once again into a super-dense
'singularity' and that another Universe will be born in
another Big Bang – a cycle that could be repeated for ever?

In the early 1960's, Hubble's law of the redshift-distance relationship for galaxies was almost completely accepted by astronomers, and the evidence accumulating from studies of distant objects was beginning to tip the balance slightly in favour of a Universe that is evolving – changing with time. The Big Bang began to look stronger as a theory than the Steady State, although the case was far from proven.

Then, in 1963, there came a bombshell. A combination of radio astronomy and optical techniques revealed a completely new and unexpected kind of object in the Universe, the quasars. And these objects had such remarkable properties that, for a time, it seemed that their behaviour cast doubt on the validity of Hubble's law, removing the very foundation stone of the science of observational cosmology. For a time astronomers actually had to ask themselves not whether either Steady State or Big Bang theory was a better description of the Universe, but whether they really knew anything fundamental about the workings of the Universe at all.

### The Coming of the Quasar

By 1962, radio astronomy had already begun to change our pictures of the Universe, revealing that many distant galaxies are intense emitters of radio energy, and must be associated with powerfully energetic astrophysical processes. But it remained very difficult to identify the exact optical counterparts of many radio sources, because the radio techniques then available gave only a rough indication of the direction of such a source, identifying a patch of sky in which there might be dozens or hundreds of visible galaxies, any one of which might be the source of the radio emission. (Today, more precise radio observations are possible, see Chapter 9).

The key to the discovery of quasars was the Moon, which astronomers use to locate distant radio sources. In effect, the radio astronomers used the whole of the Earth-Moon system, 250,000 miles long, as their 'sighting tube'. During its orbit around the Earth, the Moon passes in front of all the objects on a narrow band of sky. It happens that one interesting and unidentified powerful radio source lay in this band, and that the Moon would pass in front of it during 1962, blocking off its radio noise temporarily. By observing the source with a radio telescope and waiting for the moment of cutoff, the astronomers knew that the source must lie somewhere along the 'front' edge of the Moon's disc at that second; when the signal reappeared, the source must be somewhere along the back edge of the disc of the moving Moon. Observers could time the gap between the signal's 'off' and 'on'; they also knew the time it took the various parts of the moon's disc to pass – the lunar equator hides stars for longer than the poles do. By fitting one measurement to the other, two particular spots could be identified – where the 'front' and 'back' arcs crossed. As it happened, only one of these points marked the site of a visible optical source.

The observers thus identified the optical counterpart of the radio object known as 3C 273. But to their surprise this turned out to be not a galaxy of stars, but a single star-like object.

This was puzzling enough, for it suggested that some stars inside our own Galaxy might be strong sources of radio noise. But worse was in store. When the redshift of 3C 273 was measured, it was found to be far too big to fit into any known pattern of star movements; and, if interpreted in line with Hubble's law, the redshift placed the object well outside the Milky Way, at galactic distances. 3C 273 didn't seem to be a star at all, but rather a star-like bright source as far away as a galaxy – hence the name, 'quasi-stellar' source, from which 'quasar' is derived.

To be so small that it looked like a star, yet so bright that it was visible from as far away as a galaxy, 3C 273, and the other quasars that began to be identified in the wake of this discovery, must produce as much energy as a galaxy but all from within a region no bigger than the nucleus of a 'normal' galaxy like our own.

As more and more quasars were discovered, many with redshifts so big that they made 3C 273 seem like a near neighbour, the problem got worse. Could there really be objects in the Universe which were much smaller than galaxies but radiated as much power as all of the stars of a large galaxy put together? Naturally, astronomers began to look for alternative explanations. But these were just as worrying. Could quasars be fairly ordinary stars, much closer to home? That meant that Hubble's law might not apply to them, and once doubts were cast on Hubble's law as applied to quasars, there seemed to be room to doubt the law as universally applicable to galaxies.

Some astronomers argued that quasars might be 'local' objects, shot out from the centre of our own Galaxy in some vast explosion hurtling away from us so fast that their light is redshifted. But, if so, where was the source of all that energy?

Others argued that the redshifts might be nothing to do with velocity at all, but might really be produced by very strong gravitational fields, in line with Einstein's theory, which tells us that light struggling away from a very massive object will be redshifted in the same way as light from a rapidly receding object.

It took a good ten years for the debate to be resolved. Throughout most of the 1960's, there was no accepted picture of the Universe which could fit quasars into the same framework as galaxies, and while most astronomers clung to the Hubble law for galaxies, with its implications of an expanding Universe, the heretics could argue from a strong position that whatever strange process produced quasar redshifts also produced galaxy redshifts, and that the Universe might after all be static, as Einstein once thought.

It would simply be confusing to describe here all the different ideas that were thrashed out over the 15 years following 1963; but it is important to realize how the dramatic shock of the discovery of quasars opened up astronomical thinking.

It is really only since 1963 that we have come to appreciate just how violently active a place the Universe is. The discoveries of pulsars and X-ray sources (see Chapter 3) encouraged theorists to develop their ideas of what happens to matter when it is compressed to high densities. What they came up with – the modern understanding of black holes – carries across into the study of quasars to explain the origin of their energy.

Quasars come in a variety of shapes and sizes, sometimes with a radio structure spreading across millions of light years, but all of them deriving their energy from some tiny central source, which may produce bursts of energy up to ten thousand times the energy of all the stars in our Milky Way Galaxy put together. From Einstein's most famous equation, $E = mc^2$, which tells us how much energy could be obtained by converting *all* of a mass into energy ($c$ is the speed of light); and assuming that only a fraction of this mass-energy is being liberated in quasars, we know that the *total* mass involved must be much more than a million times the mass of the Sun. Yet it is typically squeezed within a volume of space no larger across than our Solar System – the size of the largest stars. Such an object can only be a giant black hole. But black holes, in theory, *absorb* radiation; they cannot shine as the brightest phenomena in the Universe.

But there is no paradox. It is not the quasar that shines; it is the matter around it. A quasar is a supermassive black hole, formed as the result of a lot of matter collapsing together in one place. Gravity pulls the matter into a compact state, and as more matter piles on top this becomes more and more compressed, so that the gravity field surrounding the object becomes stronger and stronger. Eventually, nothing, not even light, can escape its grip. Inside the black hole, the mass must, according to the equations, continue to collapse into a mathematical point, the mirror image of the explosion outwards from a mathematical point that is at the heart of the Big Bang

model – but that bizarre pattern of events is beyond our observing. All we can know about directly is what is happening at the edge of the black hole, where any passing material will quickly be swallowed up in the maelstrom, a kind of gravitational whirlpool in space.

If the 'hole' were very big, then matter could quietly slip into it without a large release of energy. But if the mass of several million Suns is compressed into the space of the Solar System at the heart of a galaxy like the Milky Way, then there will be a massive pileup of material – gas, dust and even whole stars – sucked in by the intense gravity field but unable to squeeze immediately into the tight 'throat' funnelling down into the hole.

With the effects of rotation and magnetic fields added in, we now have a very efficient cosmic energy source – a central black hole surrounded by a swirling mass of material which is constantly being fed from outside and heated up by collision; as much as 20% of this whirlpool mass can be turned into radiation at wavelengths that span the electromagnetic spectrum – and into escaping energetic particles (cosmic rays). This is how a quasar can shine for a hundred million years or more while devouring the heart of a galaxy.

Such a process also explains the spread of radio emission into a typical 'double lobe' pattern on either side of a quasar or radio galaxy. As the central mass spins, winding its magnetic fields around itself, it spreads a blanket of infalling material in the plane of rotation, like the rings of Saturn but on a vastly greater scale. The heat generated in this mass produces energy which powers the fast particles of cosmic rays. Some of the energetic particles escape along the 'lines of least resistance' at the poles, producing jets pushing out on either side of the central black hole.

The whole picture hangs together in a thoroughly satisfactory way, given the reality of black holes and the efficiency with which gravity

According to current thinking, the presence of all matter curves space-time to create gravity. A black hole distorts the fabric of space-time to such an extent that beyond a certain point – the 'event horizon' – electromagnetic radiation itself (blue arrows) is locked in by the intense gravitational field, which steadily consumes nearby matter (red arrows). What happens to matter beyond the event horizon – at the hole itself – we can never directly know. But the presence of such objects can be deduced by the radiant energy of matter as it crowds into the lip of the hole.

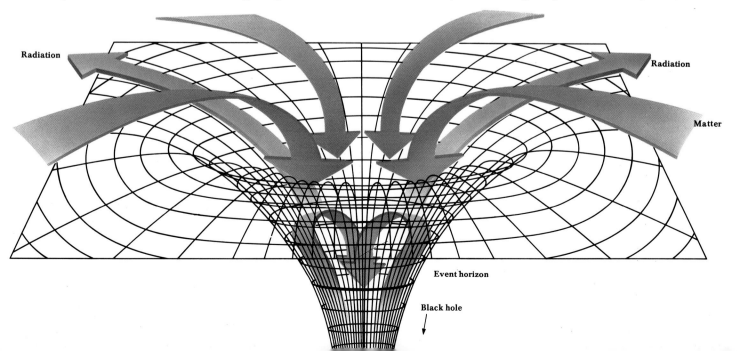

Radiation  Radiation

Matter

Event horizon

Black hole

Since the discovery of quasar 3C 273 in 1962, the speeds – and distances – of remote objects have stretched out considerably. This graph shows one hydrogen absorbtion line, known as Lyman Alpha, redshifted by two objects. The scale, in Angstrom units (1,000's) shows observed frequencies. The top line is 'static' – it represents absorbtion in our own atmosphere. The other two lines, which record radiation emitted at precisely the same frequency, have been redshifted to much higher frequencies. The second line is of 3C 273 and indicates a recession speed of 15 per cent of the velocity of light. The bottom line is of quasar oH 471, receding from us at 80 per cent of the speed of light.

can turn a million or so solar masses into energy Of course, the energies are like nothing we experience on Earth, or even in our own Milky Way Galaxy; but why should they be? The Universe is not only much bigger than we can really imagine, it is unimaginably more violent than anything conceived of only a few years ago.

The energetic activity harks right back to the Big Bang; for, with the energy source of quasars no longer a mystery, there is no need to invoke any bizarre explanations for their redshifts, and we can once again have faith in the fact of universal expansion and in the validity of Hubble's law.

Now, quasars provide a bonus for the cosmologist. Because they are so bright, many of them can be detected from much further away across the Universe than any ordinary galaxy; with the aid of redshift measurements, they provide us with the best cosmological probes of all, and scope to test the evolutionary theory of the Universe against the rival Steady State.

The techniques mentioned at the end of Chapter 7 depend on counting the number of sources we detect at different redshifts, equivalent to counting the numbers active at different times through the history of the Universe. Some evidence in this direction comes from 'ordinary' radio sources, but the best comes from quasars. Quasars are not, as it turns out, evenly distributed along the redshift scale. There are more fast-moving ones than slower ones. Since the faster moving ones are further away and older, the source counts show that there were more quasars long ago than there are now.

This evidence is on its own almost enough to settle the issue, but there is still a loophole. Can anyone *prove* that the chain of logic by which we reason out cosmic distances does not contain a flaw? Can we be sure that our instruments are not fooling us in some way, and that the source counts are not being 'biased' by some kind of consistent error?

There is one piece of evidence which seems to have settled the issue once and for all. It is the most remarkable scientific observation made in modern times, equalled in philosophical importance only by Olbers' paradox. This observation too was made by radio astronomers, and it was again a breakthrough of the 1960's, although with hindsight it is difficult to understand why it wasn't made at least ten years earlier. The discovery was of cosmic radiation that permeates all of space and is a distant cosmic echo of the Big Bang itself. With that as evidence, who could remain a Steady Stater?

### The Echo of the Big Bang
Back in the 1940's, many theorists were grappling with the puzzle of just what the Universe must have been like in the period immediately after the Big Bang. In particular, they were eager to explain how the Universe ended up containing the observed proportion of basic material, about 80% hydrogen and 20% helium, from which stars and galaxies condensed, and which has

since been partly built up into other elements through nucleosynthesis inside stars (see Chapter 3 for details).

Because we have a good idea of how much helium was 'made' from hydrogen when the Universe was young, and because we know a good deal about nuclear reactions and the conditions required to allow this kind of fusion, by 1950 it was possible to get a rough idea of what happened to the Universe as it expanded away from the initial singularity. In particular, it was clear that unless the Universe had contained some inhibiting factor during the first minutes of its life, nuclear reactions would have proceeded so rapidly that many heavy elements would have been produced right away, leaving very little hydrogen left over for stars as we know them ever to get started. The inhibiting factor was intense electromagnetic radiation.

The proportion of helium actually made tells us how intense this inhibiting background radiation was early in the history of the Universe; and, allowing for the expansion of the Universe since the beginning, that makes it possible for astronomers to predict that the Universe today should still be full of radiation, but in a much weaker form. An image often used to explain the problem is that of a box full of gas at high pressure, a box that has expanded hugely. However large the box becomes, it will still contain the same amount of gas, but spread ever more thinly, at ever lower pressures. Radiation's equivalent of pressure, in this context, is temperature. The expansion of the Universe spreads the radiation more and more thinly and the temperature drops. The prediction, even from the first calculations of this kind, was that there should still be a cosmic background radiation, with a temperature equivalent to a few degrees above absolute zero ($-273.2°C$ is absolute zero, the theoretical point at which all molecular motion ceases.) Weak, to be sure; but radiation that ought to be easily detectable with simple radio telescopes, such as were available in the 1950's, at microwave frequencies. Yet no one even looked for such radiation then, and when it was found in the mid-1960's the discovery was by accident – the predictions had been largely forgotten and ignored.

In 1978, though, that discovery turned out to have been a particularly lucky accident for Arno Penzias and Robert Wilson, who made the first observations of the cosmic background, since they received the Nobel Prize for their work. Their story begins in 1964, when as young radio astronomers working at the Bell Telephone Laboratories in New Jersey they happened to have access to an unusually sensitive radio antenna-receiver system, designed and built for communications using the Echo satellite. Penzias and Wilson were interested in measuring the background radio noise coming from the Milky Way, and were considerably puzzled when they found, in 1964, a faint radio hiss, a universal background, coming from all directions in space.

# Einstein's Universe: Strange but True

Albert Einstein, whose ideas were published over 60 years ago, is still regarded as the greatest theoretical physicist of this century, and will probably remain so until well into the next one. His theories of relativity – the Special Theory (1905) and the General Theory (1916) – were so complex, and so revolutionary that their conclusions took decades to assimilate and are still debated.

The Special Theory is founded on the principle that the velocity of light is the same for all observers, no matter what their motion relative to each other. The velocity of light is thus the ultimate speed limit of the Universe. At speeds approaching that of light, time begins to run more slowly, objects contract along their own length and they become increasingly massive. In Einstein's famous 'Twins Paradox', one fast-moving space-travelling twin would age more slowly that his stay-at-home brother. Einstein also showed the possibility of changing matter into energy, a theory summarized by his famous equation $E = mc^2$ (i.e. the energy of an object is equated with its mass times the square of the speed of light).

These bizarre predictions, which demand we abandon our common-sense notions of space and time, have been confirmed in many ways by experiment. For instance, orbiting cesium clocks, accurate to one part in many millions, have been measured running slower than equivalent clocks on Earth.

The General Theory extends relativity to gravitational fields. Einstein concluded that the presence of matter distorts space and time – space-time must be regarded together as curved. No-one can visualize curved space but the distortion is something like the 'distortion' of the Earth's surface when mapped in Mercator projection. The two-dimensional flatness of the map's surface represents a 'flatness' curved in three dimensions. In the same way, according to Einstein, we normally view the Universe as a three-dimensional map of a four-dimensional reality. The presence of matter creates additional local curvitures (as mountains do on the Earth's surface). To make another comparison, a star is like a metal ball placed on a rubber sheet. Other objects like stars and planets tend to 'roll' towards it down a kind of gravitational slope created by its presence in the Universe.

The effects of space curviture can be calculated and checked. A massive star's gravitational field should slightly reduce the energy of radiation leaving it. Starlight is bent as it passes the Sun. Mercury's eccentric orbit should revolve at a different speed than that predicted by Newton. Such effects have been verified.

Exactly *how* the Universe is curved, no-one yet knows. If it is positively curved, it may be finite in size, yet have no boundary, like the surface of the Earth. If a person could travel in an apparently straight line far enough, he would find himself returning to the point from which he began. Alternatively, it may be negatively curved, in which light follows an open path, like a parabola.

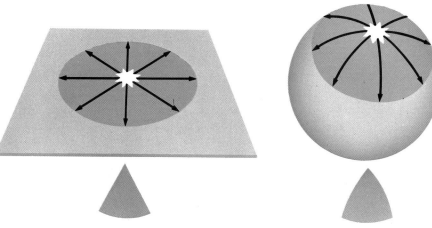

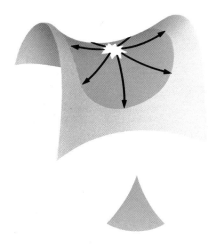

**Flat Universe**     **Positively-curved Universe**     **Negatively-curved Universe**

Arno Penzias (*left*) and Robert Wilson stand on the antenna with which they measured the temperature of the Universe in 1965 – a temperature of 2.7°K (−454.8°F), consistent with a cosmic 'Big Bang' 10,000–20,000 million years ago and the consequent expansion of the Universe since.

Over several months, they repeated the observations and tested their equipment to make sure the observations were genuine and not the result of faults in the system, but they ended up with the same result. It was only in 1965 that Penzias heard, at second or third hand, about new work by Princeton University theorist P.J.E. Peebles, who had done some new calculations predicting that the Big Bang origin of the Universe should have left a residue of radio noise detectable today as radiation with a temperature of a few degrees – no more than 10° above absolute, which is usually referred to as 0° on the Kelvin scale, named after the British physicist Lord Kelvin.

This immediately explained the observations, and caused enormous excitement as the news spread throughout the world's astronomical community, not least because the detection of the 'echo of the Big Bang' firmly nailed the lid on the coffin of the Steady State theory. But among all those excited astronomers there was one team that must have felt a deep sense of frustration. For two students, P.G. Roll and D.T. Wilkinson, had already been working at Princeton under the guidance of an older hand, Robert Dicke, to build a small radio telescope intended to test Peebles' prediction. They did this, confirmed the discovery made by Penzias and Wilson, and were in fact so hot on their heels that their publication of their own first observations of the background radiation appeared alongside the announcement of the discovery, the scientific paper by Penzias and Wilson, in the specialist *Astrophysical Journal*.

It seems more than a little harsh that the team which actually predicted the existence of the background, then went out and found it and explained exactly what they had found ended up with only a footnote in the history books, while the team that found the background by accident

and couldn't explain it at all until prompted by others ended up with world acclaim and recognition from the Nobel Committee.

But there is another ironic footnote to the story. One of the first groups to make any prediction of the nature of the cosmic background radiation had been working on the problem as long ago as 1946. Their work, however, wasn't known even to the Princeton group of the mid 1960's – which is surprising, since one of those theorists of 1946 was the same Robert Dicke who was on the Princeton team in 1964! Steven Weinberg, discussing this curious tale in his book *The First Three Minutes*, comments: 'This is often the way it is in physics – our mistake is not that we take our theories too seriously, but that we do not take them seriously enough. It is always hard to realize that these numbers and equations we play with at our desks have something to do with the real world.' Had any radio astronomer of the 1950's accepted the existing Big Bang calculations as really the best description of the Universe, he could have made the relatively simple observations involved with equipment existing at that time, killing the Steady State theory before its battle with the Big Bang cosmology really got under way – and, incidentally, writing his own name into the history books.

Since 1964, observations of the background radiation have been improved until its 'temperature' can be set very accurately at 2.7K; that in turn provides a fine tuning so that cosmologists can, in their mind's eye, wind the clock back 15,000–20,000 million years to interpret very accurately what happened to the Universe, in terms of the all-important balance between matter and radiation, in the first few minutes after the Big Bang.

## A Brief History of the Universe

We have become accustomed in this book to immense spaces of times. Now we have to slow the clock right down, for the story of that first immense explosion takes longer to tell than the events themselves.

We can pick up the story from the time just after 'the beginning' when the temperature of the fireball was a million million degrees. At such temperatures, particles of matter such as protons and electrons interacted continuously with their sub-atomic mirror images – their 'anti-particle' equivalents – and with the high temperature (that is, highly energetic) radiation background in a maelstrom of reactions. All the while, as the seconds passed, radiation was being turned into matter and matter into radiation.

As the Universe expanded, as seconds became minutes, the radiation cooled (became less energetic), and things became a little more orderly. The mass-energy increasingly stayed locked up in material particles, with the more massive particles settling out from the maelstrom first, as expansion continued and the energy density of the background radiation declined. At

100,000 million degrees, the protons and neutrons destined to make up virtually all the matter in the Universe as we know it had stabilized, but electrons and positrons continued to interact with the fading radiation left over from the primeval fireball.

At 1,000 million degrees, the weakening background radiation lost its capacity to make electron-positron pairs, and the left-over matter for the Universe as we now see it was fully settled. At about this temperature, too, the background allowed some of the protons and neutrons to 'cook' into helium, just like the nuclear fusion which operates in the Sun today.

Everything from the 'beginning' to the end of the era of nucleosynthesis – from a million million degrees down to 1,000 million degrees – took just over three minutes (hence the title of Weinberg's book). From then on, the time-scale began to stretch out.

The thousands of millions of years since the fireball have been taken up with the birth of stars, the condensation of galaxies, appearance of solar systems and – almost certainly not just on Earth – the development of life.

After the turmoil of the first million years or so following the Big Bang, the Universe was left full of swirling clouds of material, chiefly hydrogen and helium gas. At this time, and ever since, two conflicting forces were acting on the material. One was the universal expansion, tending to stretch the gas thinner and spread molecules and atoms apart from one another; the other was gravity, tending to hold things together so that if once a group of atoms combined in a swirling eddy they would attract others and grow into a bigger and bigger irregularity.

In a perfectly uniform Universe, there is no way in which large concentrations of matter could ever occur; indeed, it is quite difficult to explain how concentrations of matter as big as the galaxies can have formed in the time since the Big Bang, no more than 20,000 million years. Something in the initial fireball, it seems, must have produced density fluctuations, so that after the first million years the Universe was already 'lumpy' enough for galaxies to be able to grow. How this happened is not known; but it is straightforward to calculate how a galaxy would form from a 'proto-galaxy', one of these clouds of gas, held together by gravity, and containing enough material to form thousands of millions of suns.

The standard picture of galaxy formation envisages the gas collapsing first into a roughly elliptical shape under the influence of gravity, with stars forming out of irregularities in the collapsing cloud. At first, large hot stars, composed just of hydrogen and helium, will form in the young galaxy, run through their life cycles quickly and explode, scattering heavier elements into the interstellar medium. From these materials, 'second generation' stars can form, stars like our Sun, which have enough heavy elements for the leftover debris of star formation to produce planets and even life.

Heavy elements mix with gas into a dusty band and the whole galaxy settles into a spinning spiral – a picture that neatly explains how the Milky Way evolved, and why the stars of the halo are old and lack metals, while those of the disc are young and metal-rich. It is harder to explain why some galaxies stay as ellipticals, although it may be that this is related to how fast

The Sombrero Hat Galaxy (M 104; NGC 4594) is a spiral some 10 times the mass of our own Galaxy. It lies 41 million light-years away in Virgo. It may be a 'young' galaxy with a core of hot stars and a bank of gas from which second generation stars will later form.

the proto-galaxy was rotating, and to the power of the initial stellar explosions. Without much rotation to slow the collapse, very large early stars might form in the nucleus of an elliptical galaxy, then explode with such violence that they sweep gas and dust right out of the galaxy, leaving no scope for second generation stars to form. This would also, incidentally, leave a large mass at the nucleus, ready to coalesce into a black hole.

Although the pieces of this picture hang together fairly well, there are astronomers who do not believe that 'irregularities' as big as galaxies could have grown out of the initial fireball at all, unless something much more 'lumpy' was present to start with. They argue that from the very beginning there may have been compact, massive irregularities in the Universe, 'seeds' on which the material of the galaxies could condense by gravitational attraction.

The story is far from complete and astronomers are still puzzling over the differences between spiral and elliptical galaxies, and over the origin and evolution of both.

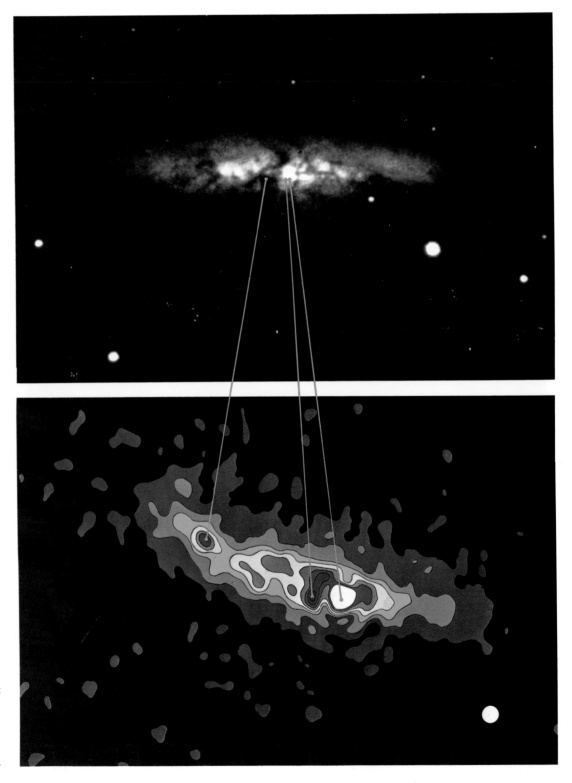

The irregular galaxy M82 (NGC 3034) consists mainly of billowing clouds of exploding gas and dust. Its central region is an active radio source (below, with colours showing increasing intensity towards the white centre). The galaxy is the nearest radio source to us, at 10 million light-years. What caused the explosion, which occurred between one and 10 million years ago, is not known.

## The Violent Galaxies

They have a mass of extraordinary objects to study, all of which could provide additional pieces in the puzzle of galactic evolution. Quasars are not the only violent active phenomena in the Universe. There are, for instance, exploding galaxies of many different kinds. For a while, astrophysicists were faced with the daunting prospect of having to find a different explanation for every kind of exploding galaxy. But today a coherent picture has emerged, with quasars seen as simply the most extreme version of a kind of activity that spreads right across the scale down to such ordinary galaxies as our own Milky Way. *All* galaxies, it now seems, may be active from time to time during their lives, and *all* galaxies including our own, may harbour black holes at their centres.

The classic examples of exploding galaxies are called Seyfert galaxies, which show activity somewhere in the middle range between a galaxy like the Milky Way and a quasar. This type of galaxy was first identified in 1943 by Carl Seyfert, after whom they are named. He found that about one per cent of all spirals have very insignificant spiral arms but very bright central regions (nuclei). Such systems have now been studied across the spectrum – from X-rays through ultraviolet and optical light into the infrared and radio regions; the observations show that the radiation of the bright nucleus is produced by hot gas in violent motion, with a central 'condensation' concentrating the mass of millions of suns in a small volume. This, of course, is just the kind of system now thought to provide the energy for quasars, the difference being that it is operating on a slightly smaller scale and that we can see (just) the spiral arms of the surrounding galaxy.

On either side of the Seyfert phenomenon in the chain of energies, there are active but apparently more 'normal' galaxies (intermediate between Seyferts and galaxies like our own), and even more compact and energetic objects, the N-galaxies (intermediate between Seyferts and quasars). As far as physical characteristics are concerned, there seems to be a continuous gradation from the most compact energy sources, quasars, right through to quiet galaxies. It may very well be that the differences are simply due to the different amount of matter that is present in the central black hole in each case. As more and more material is swallowed up in the nucleus, any particular energetic galaxy may slowly be converted into the next step up the chain.

During its active life, a system might undergo several spasms of activity, rather than shining brightly all the time. It may swallow matter up in bursts, then 'rest' while more matter falls in towards the black hole. Instead of one per cent of spirals being Seyferts, it is more likely that *all* spiral galaxies are Seyferts for one per cent of their lifetimes. Even our Milky Way Galaxy shows signs of a peculiar source at the galactic centre, modest by some standards, but perhaps a black hole with a mass of five million suns, sufficient to produce energetic activity from time to time, although not quite enough for our Galaxy ever to have been a quasar.

Eventually, though, any black hole of this kind will sweep the central regions of its parent galaxy clean, so that very little material is left to fall in, although stars may continue happily in their orbits far away from the nucleus. The quasar or Seyfert activity will fade away, and what we think of as an 'ordinary' galaxy will be left. But the fading away may take millions of years – in the meantime forming giant radio sources, the last pieces in the puzzle of galactic evolution.

Radio galaxies were first identified in the 1950's, when radio observing techniques became accurate enough to pinpoint the optical counterparts of some of the many strong celestial radio sources that were already known then. One of the first identifications made was of the brightest radio source in the direction of Cygnus, dubbed Cygnus A. This is the second brightest object in the sky at radio wavelengths, and it had seemed natural to guess that it must be fairly nearby. Yet the identification, made in 1954, showed that Cygnus A is actually associated with a galaxy, shown by Hubble's law to be almost 650 million light years away from us. For the radio energy of the galaxy to be so bright at such a distance, it must be 10 million times more powerful, as a radio source, than a galaxy like Andromeda or the Milky Way.

Progress in identifying radio galaxies was slow, and by 1970 only a couple of hundred positive identifications of radio sources with galaxies had been made; but now the figure is well above 1,000, and the broad features of this class of cosmic phenomenon can be discerned. It turns out that Cygnus A, the first to be identified, is indeed typical of its kind.

Once again, radio galaxies typically show signs of active central nuclei; the most powerful radio galaxies are generally associated with giant elliptical galaxies; and, most intriguingly and significantly of all, most powerful radio galaxies show a characteristic 'double lobe' pattern, with the most intense radio noise coming from *either side* of the central galaxy, with only a weak radio emission from the nucleus itself.

There is very little doubt that these powerful radio sources are the result of gigantic explosions in the nuclei of the galaxies with which they are associated. Energetic charged particles are squirted out in opposite directions, and produce radio emission as they interact with the magnetic fields around the galaxy. But these moving particles can only move out at speeds less than the speed of light, and in the most extreme case, the giant radio source 3C 236, the ends of the radio lobes are 20 million light years apart. In other words, the explosion which produced the radio source we see now occurred at least 10 million years ago, since it must take that long for the material in each lobe to get where it is today. To

Centaurus A (NGC 5128), a giant radio galaxy in the southern hemisphere some 16 million light years away, emits radio waves from two lobes about 100 times the size of the visible galaxy.

put this in perspective, the distances involved are roughly 10 times the distance to our near neighbour, the Andromeda galaxy.

On the other hand, the fact that no bigger sources are found also tells us something about the activity of galactic nuclei, for it seems that these giant explosions never proceed for more than a few tens of millions of years without dying away, and certainly not for anything like the lifetime of a galaxy, which is around 10 *thousand* million years. What happens when they do fade away?

We can get a very clear picture of what seems to be a once powerful radio galaxy now on its last legs in our near neighbour Centaurus A, a mere 16 million light years away. This has very large radio components, stretching across two million light years, but is very weak as radio galaxies go, with just one-thousandth the power of Cygnus A. As well as the giant clouds of radio emission there are two smaller clouds just emerging from the edge of the galaxy, and the central radio source provides fully a fifth of the total radio noise, a high proportion by radio galaxy standards. So there is some evidence that after a huge outburst millions of years ago Centaurus A has had a second hiccup of activity on a lesser scale. This is just the kind of behaviour that would be associated with a once active central black hole, initially swamped by matter pouring onto it, which has now cleared most of the space around itself and is kept modestly active by a last trickle of remaining mass being swept up as the hole moves into a quiet old age. In all probability, Centaurus A is the nearest massive black hole still showing the effects of this accretion. Once, the galaxy must have been as energetic as Cygnus A is today.

Many quasars are also powerful radio sources, and again they show a characteristic preference for the double-lobed structure, with trails of radio emitting material shot out and spreading across the light years. The relationship with radio galaxies seems clear, and there is no need to invoke any different processes to explain these superficially different phenomena after all. Most quasars must now be dead, their central black holes starved of matter to swallow up and turn into energy, so that they sit quietly at the centres of galaxies. Some, like Centaurus A, are dying; some holes never quite got enough mass together to become quasars, but have had an active life as the powerhouses of Seyferts, N-galaxies and the like. The Universe is a violent place, but less violent than it was, and getting less violent still as time goes by. Will it fade away into an ever expanding sea of ever quieter galaxies, with the lights of the stars and galaxies going out one by one until all that is left is a void filled with a scattering of black holes? Or is some more interesting fate in prospect?

## From Big Bang to What?

We have already taken a look at the beginning of the Universe. What will be its end? And why is it constituted the way it is? The answers may bring us full circle and tell us what was there before the 'beginning', and what there may be after the 'end'.

The crucial factor turns out to be how much matter there is in the whole Universe. For, just as galaxies are combined by the attractive force of gravity, the total mass of the whole Universe provides an insistent tug on every galaxy and cluster of galaxies within the Universe. As the Universe expands outward from the initial explosion of the Big Bang, the gravitational tug gradually decreases. But it is always present, and if there is enough matter, eventually gravity will overcome the expansion, and the Universe will slow to a halt. The whole drama will then be played out in reverse. The Universe will collapse faster and faster under the overwhelming pull of gravity until it is squashed into another fireball, perhaps then to bounce back out again.

On the other hand, without enough mass the expansion can never be reversed, and the spread of matter in the Universe will indeed get thinner and thinner forever as the Universe ages.

By measuring the amount of matter – the number of galaxies – in the volume of space we can see, cosmologists get an indication of what the density of matter is throughout the Universe. Intriguingly, but annoyingly, the best estimates we have are that the Universe is balanced on a knife edge. As far as we can tell, there is either just enough matter to 'close' the Universe and make it, eventually, collapse; or there is not quite enough, so that it will expand forever. More as an act of faith than anything, most astronomers at present seem to prefer the continuously expanding model. They could well be wrong, for one thing is clear: we are not going to find any *less* matter in the Universe, while there

might well be material we don't yet know about – cold gas between the galaxies, for instance, or black holes at present undetected. So even now the two versions of the fate of the Universe should be given equally serious considerations.

This is just as well for the theorists, since there is really nothing more to be said of the continuously expanding – 'open' – Universe than that it will expand forever and die. If this is the ultimate destiny of the Universe, we are very lucky to be around at a time when it is relatively young and active, with so many interesting phenomena to observe.

But the closed universe model is much more interesting. Apart from anything else, it removes the puzzle of the 'beginning'. For now we can say that before the Big Bang there was another cycle of expansion and collapse, and that these oscillations have been going on for ever, and will go on for ever.

Once we start to ask these deeper questions about the nature of the Universe – philosophical questions, if you like – some fascinating puzzles emerge, not least being the way the Universe is just right for life as we know it. Our surroundings really are pretty unlikely, in terms of the standard rules of physics. We know from physical experiments that the most likely state of any system is one of uniformity, which would correspond to a random spread of matter across space, with no clumps of galaxies, stars and planets (let alone life) to provide order among the chaos. In physics terminology, the Universe contains a lot of information – it is a complicated place. And that is very unlikely, if we are dealing with a random state.

Does that mean that the Universe is *not* the result of chance? Have cosmologists rediscovered God? Maybe – if this Universe is a one-off. But if it is cyclic, it is possible to envisage not just one Universe arising out of the initial conditions of the Big Bang, but many alternative possible Universes, depending on just which way many, random (chance) processes turn out as the Universe expands. It might be that all of the infinite variety of possible Universes get a chance to develop into reality. In the infinite chain of cycles stretching off into the past and future, everything that can possibly happen does happen. 'Our' Universe is a very unlikely development, perhaps, but it must happen sometime. We are here to see it because life can develop in this kind of unlikely universe, with the kinds of stars and galaxies and planets that actually exist.

This must be the ultimate step in the long process of setting man in the context of Creation. Less than 500 years ago, he seemed its very centre. Over the centuries, we have discovered that the Earth orbits the Sun, not the other way around; that the Sun is just one insignificant star in the Milky Way Galaxy; that the Milky Way is but one modest galaxy with no special place in the Universe; and now that the Universe as we know it is, perhaps, just one Universe in an infinite number.

Two possible versions of the origin and fate of the Universe show the fine difference there seems to be at present between cyclical Big Bangs and infinite dispersion. The only way to find which (if either) is correct is to assess accurately the amount of matter in the Universe.

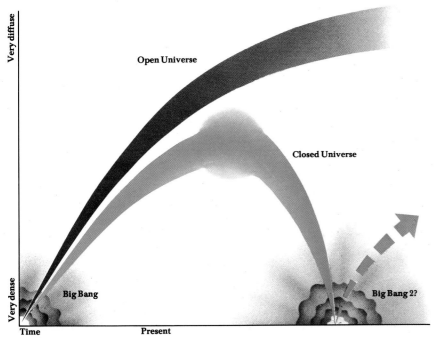

Very diffuse

Open Universe

Closed Universe

Very dense

Big Bang

Big Bang 2?

Time    Present

# 9/THE QUEST FOR KNOWLEDGE

As the history of astronomy itself shows, the difficulties of discovering anything firm about the Universe are immense. All we have to go on is the information conveyed by light. Until very recently, mankind could do nothing but observe – and that poorly, immersed as we are at the bottom of a sea of hazy atmosphere. Over the last few decades, however, scientists have been able to undertake research with astonishing new tools – radio-telescopes that analyse invisible regions of the spectrum (like the dish at left at the Mullard Observatory, Cambridge), devices that can count the basic constituents of light and information from satellites outside the atmosphere. These – and others – are the techniques that have sparked the current revolution in astronomy.

In many ways astronomy – unlike other sciences – has changed little with the centuries. There have been vast improvements in techniques; but these have been refinements rather than differences in actual approach. Until the last decade, astronomy has been almost exclusively a purely observational science. Experiment, the lynch-pin of most other sciences, has little place in astronomy. An astronomer cannot dissect a star: all he can do is observe its radiation, and then attempt to say something of its motions and structure. This has given astronomy an extraordinary historical continuity.

Astronomy's beginnings will always remain shrouded in the mists of antiquity. We know from the great megaliths that Man must have had an intimate and sophisticated knowledge of the workings of the cosmos at least 4,000 years in the past. Earlier perhaps than this, men living in the Near East and around the shores of the Mediterranean had begun to gaze heavenwards. Blessed with a good climate and clear skies, they soon became familiar with the starry vault. As we have learned in Chapter 4, these early astronomers joined up the stars into an imaginative array of groups and patterns which wheeled in procession across the sky from dusk to dawn, making up a huge celestial clock.

Their clock also doubled as a calendar, for it soon became obvious that changing star patterns

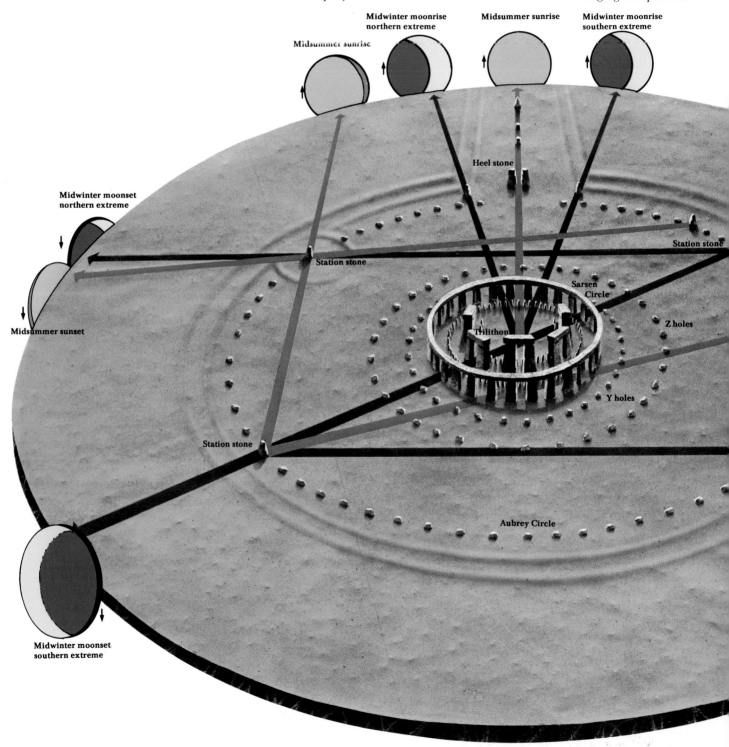

became visible as the days wore on; and the power of these first astronomers lay in their ability to predict regular earthly events from different configurations of this heavenly calendar. For example, the appearance of certain constellation patterns heralded the coming of spring and with it, the signal for the farmers to sow their crops; and the Egyptian astronomers knew that when Sirius, the brightest star in the sky, rose just after sunset, the mighty Nile would burst its banks and start its much-needed annual flood. To the uninitiated, it must have seemed almost miraculous that the stars could foretell – or even cause – events upon which Man's very survival depended. By extension, it was assumed

the stars controlled *all* human life, and astrology was born.

The early watchers noticed, too, that other bodies besides the stars inhabited the heavens. Most obvious of these were the Sun and Moon, whose extreme brilliance made them very important. There were other objects, too, which behaved in a way which set them apart. Whereas the stars stayed fixed in their constellation patterns, the five bright 'wandering stars' (called *planetes* by the Greeks, hence our modern word 'planets') shifted slowly against the starry background in a manner which was independent of the motion of the sky.

Among the first peoples to leave written ac-

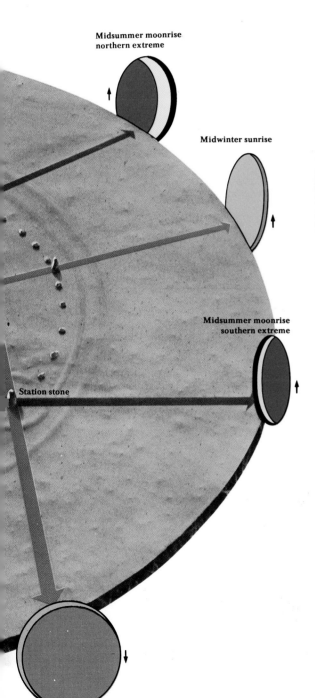

counts of their astronomical observations were the Babylonians, active around 1500 BC, who inscribed their records on clay tablets. Their interest in the heavens was not a purely objective one, as that of a modern astronomer would be: the observations they made, and their interpretation of them, were partly motivated by astrological concerns. They believed that the heavenly bodies had an influence not on individuals, but on the governing of a country and its affairs of state. In such times of political unrest, it would clearly be advantageous to see into the future. The sorts of observations which they needed for these divinations were of the movements of bodies in the sky, and of phenomena which were cyclic; but they showed little interest in recording transient events such as novae, supernovae and the appearances of bright comets. And so, from this observational bias, we see how earthly concerns were able to influence and even dictate the progress of a science.

Despite these selection effects, the Babylonian astronomers were careful and diligent observers. Watching the sky from the roofs of specially constructed towers ('towers of Babel'), they mapped the brighter stars and paid particular attention to the band of constellations along which the Sun, Moon and planets appeared to move, dividing it up into 12 sections (as we have

Stonehenge on Salisbury Plain, is an astronomical observatory of remarkable sophistication. Built in several stages from 2900 BC to 1400 BC, this neolithic computer records the sunrise, sunset, moonrise and moonset at mid-winter and mid-summer. The key alignment is that from the centre through the heel stone to mid-summer sunrise. Four 'station stones' – part of the so-called Aubrey Circle of 56 holes – mark a number of the observations. Because the Moon's orbit 'rocks' slightly, the Moon rises and sets at a slightly different place every day, in an 18.6-year cycle that is also recorded in the stones.

Midsummer moonrise northern extreme

Midwinter sunrise

Midsummer moonrise southern extreme

Station stone

Midwinter sunset

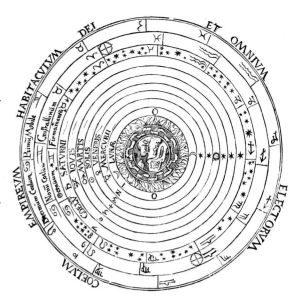

The Universe of Aristotle places the Earth at the centre with its four elements – earth, air, fire and water – surrounded by concentric spheres containing the heavenly bodies. The spheres revolve on their axes, driven by the primum mobile or 'prime mover' in the outer sphere, which medieval Christians could equate with God.

today) so that they could more easily chart the motions of the wandering bodies. They knew the length of the year, from the Sun's motion against the stars, to an astonishing accuracy of four and a half minutes, but preferred to use a calendar based on the motion of the Moon.

From these studies, the Babylonians drew up a model of the Earth and the Universe. The Earth was believed to be a flat, round disc, surrounded by water and ringed with mountains which supported the dome of the sky; the dome itself was pierced with doorways to enable the various celestial bodies to enter and leave. Completing the picture was a shell of water surrounding the dome, which was necessary to account for rain.

The Babylonians did not care to theorize as to what made the Sun, Moon and planets move, unlike the Egyptians, who flourished at about the same time. Their conception of the Universe was superficially similar to that of the Babylonians, with a flat Earth surmounted by a heavenly dome, but rather than being based on observation, it was founded on superstition. What were the heavens? The Egyptians replied that they were actually the star-spangled body of their goddess Nut, who arched in an ungainly fashion across the sky supported by the arms of the air god Shu. Why did the Sun move across the sky during the course of a day? Because it sat in a boat which made the daily voyage along Nut's body, they answered.

It is hardly surprising to learn that astronomy, as a science, did not flourish in Egypt. However, as an aid to timekeeping and calendar making it reigned supreme, for the Egyptians were first and foremost a practical race, and highly skilled in engineering. Like the Babylonians, their observations – made, at best, with crude wooden altitude-measuring instruments – were painstakingly accurate, as we know from the alignments of the pyramids and their predictions of the Nile floods.

The next race to inherit this tradition of ob-

servation and meticulous calculation were the Greeks, with whom the Egyptians and the Babylonians undoubtedly traded. But upon these foundations, they developed a discipline whose framework and conclusions were destined to guide subsequent civilizations in many fields far removed from science for the next 2,000 years.

Starting with Pythagoras, the great geometer, they replaced the concept of a flat Earth by one of a globe which remained stationary in the middle of a spherical Universe. Thus began their preoccupation with symmetry, which, with a few exceptions, coloured the interpretation of their observations for the whole time the Greek civilization flourished. Indeed, so powerful did this obsession with symmetry and perfection become under the great philosopher Plato, that those who aspired to wisdom were advised not to make any observations (of anything) at all; for the objects under observation would be but crude and misleading approximations to the real and perfect truth.

Fortunately, a number of Plato's pupils had other ideas, and sought to incorporate astronomical observations into their own teaching of philosophy. One was Eudoxus, who tried to account for the motions of the planets as the rotation of a nest of crystalline spheres centred on the Earth – the first of countless attempts to explain these complicated movements by variations on a circular theme.

Another of Plato's pupils – Aristotle – was destined to become the greatest and most influential philosopher of all time. His first appointment on leaving Plato's Academy was as tutor to the young Alexander of Macedon (later to become Alexander the Great), but he returned to Athens on the assassination of Alexander's father to set up his own school (the Lyceum). Here students flocked to hear Aristotle's teachings on philosophy, biology, medicine and astronomy. His work was startlingly original and ahead of its time, and his astronomical teachings in particular were to determine the course of Western science during a very crucial stage in its development.

On the surface, Aristotle's view of the cosmos differed little from that put forward by Pythagoras 200 years before: he too believed that the

In this representation of the Egyptian cosmos, the Earth God, Qeb, lies spanned by Shu (kneeling) representing the air and the Goddess Nut, representing the sky, arched above. The Moon (left) and the Sun (right) sail over Nut's body.

globe of the Earth remained stationary at the centre of a spherical Universe. The difference lay in Aristotle's approach to the problem. He argued his case from logical, scientific principles, relying heavily on observation. The Earth was a globe, he taught, because different constellation patterns appeared in the sky if one ventured substantially north or south of Athens; and there was the extra evidence that the Earth's shadow on the Moon looked curved during lunar eclipses. However, the Earth did not move in space, because if it did, we would expect to see the stars rising and setting in different places as time passed. All the heavenly bodies were perfect – because no imperfections had been observed. The Universe itself was symmetrical and the planets indeed travelled around the Earth, but not in as simple a manner as Eudoxus had proposed. Aristotle had to bring in some 55 homocentric spheres to clear up that problem.

Aristotle's picture of the Universe was logical, symmetrical, consistent, beautiful and, above all, scientifically reasoned. It claimed to be accurate, and therefore stimulated further observation and precise measurement. Eratosthenes, one of the first directors of the great museum and library in Alexandria (a true forerunner of a modern research institution) carefully measured the circumference of the Earth by observing the Sun's position at noon at different latitudes, getting much the same value as we do today; while Aristarchus of Samos, his contemporary, was busily measuring the distances and sizes of the Sun and Moon. Aristarchus' results were wildly inaccurate, but there was nothing wrong with the scientific principles behind his methods: it was simply that extremely precise angular measurements were required, and the requisite instrumentation did not then exist.

Aristarchus is widely remembered for his far-sighted proposals that the Earth is just one of the planets in orbit about the Sun, and that the rising and setting of the stars is simply a consequence of the daily rotation of the Earth. But his theories could not be tested by observation: they were therefore not accepted as valid.

Shaken by political upheaval, and threatened by the rise of the Roman Empire, Greek culture died around the first century AD. During the brief settled period which followed, scholarship flourished for a while, particularly in research centres like the library in Alexandria. But the mood was nostalgic, and efforts were made to incorporate the Greek teachings of the past whenever possible. One of the Alexandrian researchers was Ptolemy (Claudius Ptolemaeus), who compiled the magnificent *Almagest* – a synthesis of the astronomical and geographical knowledge of the Greek's, which incorporated many of his own ideas and observations.

He strongly supported Aristotle's picture of the Universe, believing that he had obtained additional evidence against the movement of the Earth from the compilation of his vast star catalogue. Ptolemy's was not the first. Two centuries before him, the great observer Hipparchus had drawn up a catalogue of 850 stars, arranged in the six different brightness classes (magnitudes) which we use today. By comparing his catalogue with the incomplete charts of his predecessors Hipparchus, in 130 BC, was able to establish the existence of precession. In a similar, and equally scientific fashion, Ptolemy compared his even larger catalogue (listing more than a thousand stars) with that of Hipparchus, and maintained that their similarity proved the Earth to be stationary in space.

Another of Ptolemy's innovations – and the one for which he is chiefly remembered – was in his explanation of planetary motions by a system of epicycles and deferents. On this theory, planets orbit the Earth in combination of two types of circle – a wide circle (the deferent) and individual loops (epicycles) around this mean path – thereby accounting for the strange (retrograde) motion sometimes observed. Apollonius of Alexandria had actually suggested this mechanism over 400 years earlier, but Ptolemy developed and refined it to such a degree that the *Almagest* contained tables predicting the positions of the planets for many years in the future.

Ptolemy was, in a sense, right; the planets *do* move in ways that can very nearly be described in terms of circles, except that one of the circles is the Earth's own orbit. His system was thus very accurate, but not perfect; it was also horrendously complicated. The *Almagest* was a fitting memorial to seven hundred years of Greek astronomy; and more than any other work it was responsible for laying the foundations of modern science. But such was not to be for another thousand years. Political and religious chaos soon plunged most of the civilized Western world into cultural darkness, and the flame of science was extinguished.

## The Copernican Revolution

Our story now leaps some 1,500 years, to a chilly spring in Poland where an old man lay on his deathbed. Clutched in his hands was a book destined to sow the seeds of a revolution. The year was 1543; the man, Nicolas Copernicus.

In Copernicus' day, Aristotle once again reigned supreme, but certainly not in a way that Aristotle himself would have liked. The great man's teachings had been distorted into a form which would aid and abet the power of the Church. This had come to be by a roundabout route. In the 7th century AD, the Arabs had swept across the Near East in the name of Mohammed, penetrating as far west as Egypt and Spain. Stumbling across the Greek texts which remained in the devastated Alexandrian library, they realized that some could be put to practical use, and began the task of translating them. Among the works they discovered were the *Almagest* and the teachings of Aristotle. Both helped the Arabs in constructing lunar and solar calendars for religious and civil use in formulating a picture of the Universe.

This medieval drawing of Ptolemy shows him holding an astrolabe, a device used to measure the height of stars above the horizon. His view of the Universe resembled Aristotle's in that the Earth was placed at the centre, but the motions of the planets were described in theoretical and geometrical terms as combinations of various sorts of circle. His scheme, which was complex but surprisingly accurate, was taught alongside that of Aristotle in the Middle Ages.

Copernicus, a Polish cleric, for whom astronomy was little more than a hobby, began the scientific revolution in 1543, when he suggested to place the Sun at the centre of the Solar System would provide a simpler and more accurate description of planetary motions than Ptolemy's system.

By the start of the mediaeval era, news of the Greek texts, as interpreted by the Arabs, began to filter through Spain and thence to the great monasteries and centres of learning in Northern Europe. The Crusades enabled scholars to get their hands on some Greek originals, and translation soon began. The Church – at the pinnacle of its wealth and power – was not slow to realize that the teachings of Aristotle could be re-interpreted in a Christian way. Here was a tautly argued, but non-partisan view of an Earth surrounded by a perfect, unchanging Universe. Aristotle's scientific conception was twisted into dogma to perpetuate religious power, and such an attitude stultified objective enquiry. Into this climate Copernicus was born.

Copernicus made his living as a priest, but had studied astronomy and mathematics at university and was well acquainted with Ptolemy's explanation of the planetary motions. Others, too, had been looking at these motions afresh, with a view to drawing up navigational tables for the long sea voyages of trade and exploration which had just commenced. Copernicus was dismayed at how unwieldy Ptolemy's system had become, now needing many extra epicycles to achieve reasonable predictions. How much simpler, he thought, if the Sun were to replace the Earth in the centre of the Universe; and that the Earth, along with the planets, moved around the Sun. Such a view – whatever its predictive appeal – was tantamount to heresy in the eyes of the Church.

It is still unknown whether Copernicus genuinely believed that the Earth moved, or merely regarded his suggestion as a computational device; whatever the truth, it remains that Copernicus did not publish his ideas until the year of his death. Tradition has it that the first copy of *De Revolutionibus* – the book which set forth his heliocentric theory – arrived at the dying man's bedside just in time.

Copernicus had no observational evidence to support his theory. Yet its effects were far-reaching: not only did it contradict the contem-porary dogma, but it prompted scholars to generate their own ideas at last, rather than continuing to look to the great masters of a by-gone civilization.

The one hundred years which followed were a watershed in astronomy: Copernicus' thinking was brought to fruition by the efforts of three great contemporaries – Tycho Brahe, Johannes

In this medieval painting, an astronomer takes navigational measurements – vital information with which sailors could check the date, time of day and position.

Kepler, and Galileo Galilei.

Tycho was born in Denmark only three and a half years after Copernicus died, growing up to become a colourful, if rather hot-headed figure; but his astronomical observations were quite without equal. He began his research, generously sponsored by the King of Denmark, at a time when scientific instrument manufacture was increasing in leaps and bounds. The main emphasis was on surveying instruments – mapping and town construction being of paramount importance in those financially buoyant times – but the principles were easily adapted to angular measurements in the heavens. Tycho started to record the positions of Sun, stars, Moon and planets, at first with mural quadrants, and later with portable sextants of his own design. His instruments were unprecedentedly precise, partly because their great size allowed large, easily-read scales, and also because Tycho had reasoned that each observation he made was subject to various sources of error, which he sought to track down and eliminate. In this way, Tycho made positional observations which were five times more accurate than those of Hipparchus. But because of his religious beliefs, and also because he could see no evidence for the motion of the Earth in his observations, Tycho did not support Copernicus' theory.

It remained to his assistant, Johannes Kepler, to demonstrate how overwhelmingly these supported it. Kepler, 25 years Tycho's junior, was a brilliant mathematician who believed strongly in Copernicus' ideas. He had outlined these thoughts in his popular *Mysterium Cosmographicum* where he made it clear that his motivation was not purely scientific: he believed that the distances of the planets from the Sun bore a geometrical relationship to one another as a result of their 'spheres' (orbits) being separated by one of the regular geometrical solids (cube, tetrahedron, etc).

This search for heavenly harmony led him to analyse Tycho's comprehensive planetary observations after the latter's death in 1601. Mars

was particularly puzzling, and traced a path which could be accurately predicted by neither Ptolemy nor Copernicus. Kepler hit on the answer, making the decisive step which wrenched astronomy away from all considerations of perfection: Mars did not travel around the Sun in a circle, but in an ellipse. From this discovery stemmed Kepler's laws of planetary motion, which he considered to be the ultimate confirmation of the grand design. It is somewhat ironic that we no longer remember Kepler for his harmonies, but for their bare skeleton: a powerful and elegant set of laws which allow us to determine the dynamics and scale of our Solar System.

Our final member of the 'watershed' trio – Galileo – had been appointed professor of mathematics at Padua in 1592. Italy was then in the grip of rigid pro-Aristotelian feeling, the Church authorities equating non-orthodoxy in science with unauthorized religious beliefs. Galileo made himself unpopular with the authorities because of his open, and somewhat tactless support for the Copernican theory, and he was dismissed from Pisa university as a result.

In 1609 came the news that was to hammer the final nail into the coffin of Aristotelian dogma, and the first into Galileo's own. An optician in Holland – Hans Lippershey – had patented a tube containing two lenses which made distant objects appear closer. Already, Lippershey was aware of their military application and had sold a number of these 'telescopes' to the Dutch army. Galileo was familiar with lenses and the principles of optics, as well he might be, for Venice was a centre of glass manufacture; and it was not long before he constructed his own telescope and turned it to the skies.

He was astonished at what he saw. The Moon, instead of being a perfect body as the Church preached, looked like the Earth, with seas and mountains; the inviolate Sun had spots; Venus displayed phases, showing conclusively that she orbited the Sun and not the Earth; and Jupiter was accompanied by four tiny bodies which circled the planet perpetually. All these dis-

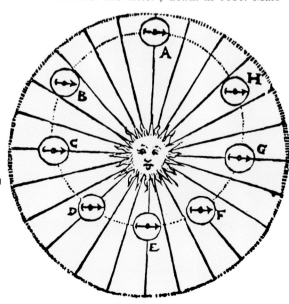

coveries were made with miniscule telescopes which would be regarded as toys today – none magnified more than twenty times.

Fired with enthusiasm for his findings, and by their support for the Copernican theory, Galileo rushed into print with a series of dangerously explicit books which succeeded in infuriating the Church authorities. Eventually, he was forced to recant under the Inquisition, but even this did not prevent him from pursuing science. For the last ten years of his life he was under virtual house arrest, and despite his growing blindness he turned his mind to the problems of moving bodies and their acceleration. Fittingly, it was the kind of work which foreshadowed that of Isaac Newton, who was born in the year that Galileo died.

England of 1642 was a very different place from Italy. Life was no longer dictated by religious observance; reading and learning flourished; exploration was the norm in all endeavours and a healthy spirit of scientific enquiry had emerged. Experimental science in particular prospered, as scientific instrument makers improved their craftsmanship dramatically; and on the theoretical side, reports of Copernicus's and Kepler's work came swiftly over from the continent and met with little resistance. News of the telescope was sweeping across Europe and astronomers and craftsmen alike tried to improve the instruments. The first telescope workshops began to be set up in London.

This heady atmosphere would have permeated the centres of learning at the time, and even quiet, introspective students like Isaac Newton – who went up to Trinity College, Cambridge, in 1661 – must have been affected. He showed little early promise, but his Professor of Mathematics, Isaac Barrow, realized that the young man had

great ability in optics. But Barrow was to see little of his student for a couple of years, for the University closed in 1665 as a precaution against the advancing plague, and Newton returned home to Lincolnshire.

Always a solitary person, Newton was glad of this sojourn. He spent the time thinking; and in two short years laid the foundations which were to dictate much of the course of science right up to the present day. He pondered on the behaviour of moving bodies, like Galileo had before him, and wondered if the stars and planets were subject to the same laws as bodies on Earth. He thought about the nature of light and colour, and came to the conclusion – after experimenting with a prism purchased from a travelling fair – that 'white' light was really made up of a mixture of colours. Had his prism been of higher quality, he would undoubtedly have discovered the principle of the spectroscope.

On returning to Cambridge – and quickly assuming the Chair of Mathematics – Newton translated his thoughts into actualities. His researches on light led him to design a telescope which collected light with a mirror, avoiding the (until then insuperable) problems of false colour and lens distortion. This 'Newtonian' reflecting telescope – which he exhibited at the newly-formed Royal Society in 1672 – was the forerunner of all the world's greatest telescopes today.

But Newton is renowned for his theory of Universal Gravitation, which tells us that the motion of a star around a galaxy can be described in exactly the same way as that of an apple falling to the ground. En route to a complete theory, Newton was forced to devise a powerful new mathematical tool, the calculus; and in 1687 he was finally persuaded to publish his results in a book entitled *Principia Mathematica Philosophiae*

## Newton and the Apple

Every schoolchild has heard the story of Newton and the apple. It sounds apochryphal. But it was a real incident, as William Stukeley in his *Memoirs of Sir Isaac Newton*, of 1752 records:

'On 15 April 1726 I paid a visit to Sir Isaac at his lodgings in Orbels buildings in Kensington, dined with him and spent the whole day with him, alone . . .

'After dinner, the weather being warm, we went into the garden and drank tea, under the shade of the apple trees, only he and myself. Amidst other discourse, he told me, he was just in the same situation, as when formerly, the notion of gravitation came into his mind. It was occasion'd by the fall of an apple, as he sat in a contemplative mood. Why should that apple always descend perpendicularly to the ground, thought he to himself. Why should it not go sideways or upwards, but constantly to the earth's centre? Assuredly, the reason is, that the earth draws it. There must be a drawing power in the matter of the earth; and it

Sir Isaac Newton.

must be in the earth's centre, not in any side of the earth. Therefore does this apple fall perpendicularly, or towards the centre. If matter thus draws matter, it must be in proportion of its quantity. Therefore the apple draws the earth, as well as the earth draws the apple. That there is a power, like that we here call gravity, which extends its self thro' the Universe.

And thus by degrees he began to apply this property of gravitation to the motion of the earth and of the heavenly bodys, to consider their distances, their magnitudes and their periodical revolutions; to find out, that this property conjointly with a progressive motion impressed on them at the beginning, perfectly solv'd their circular courses; kept the planets from falling upon one another, or dropping all together into one centre; and thus he unfolded the Universe. This was the birth of those amazing discoverys, whereby he built philosophy on a solid foundation, to the astonishment of all Europe.'

# 'The Leviathan of Parsonstown'.

Throughout the early 19th century, the world's largest telescope was Herschel's 48-inch reflector, built in 1789. But in 1845, an Irish nobleman, William Parsons, the third Earl of Rosse (*right*), built a 72-inch (183 cm) telescope (*below*), which remained the world's largest until the opening of the Mount Wilson 100-inch (254 cm) in 1917.

Lord Rosse, born in 1800, was a wealthy landowner who entered Parliament when still an Oxford undergraduate. But his chief interest was science. At the age of 27, he began to design large reflecting telescopes – those that use mirrors to gather light – which were still in an experimental stage (Herschel had never published his methods).

Rosse set about developing a new alloy of copper and tin that would take the maximum amount of polish. In 1839, he successfully made a three-foot speculum (mirror), but found that it responded too rapidly to changes in air temperature, and he moved on to more solid structures. By 1843, he had made two massive mirrors – 72 inches across

and weighing four tons apiece – which he planned to use in rotation. In 1845, the complete telescope – called the Leviathan of Parsonstown – was ready for use. Rosse mounted it at his house, Birr Castle, between two walls, 56 feet high, which acted as windbreaks for the 58-foot tube.

Although the instrument was clumsy to use, its light-gathering power was tremendous. With it, Rosse became the first to see that some of the cloudy objects known as nebulae – actually other galaxies – were spirals. He also noticed the structure of the so-called planetary nebulae (stars that have cast off their outer atmosphere) and sketched the Crab Nebula (see Chapter 5).

The telescope was used for a decade after Rosse's death in 1867, but was later dismantled.

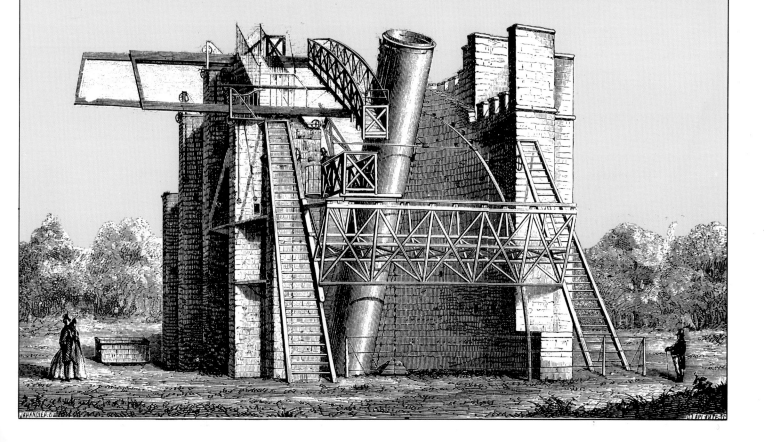

*Naturalis.* In rigorous mathematical form, Newton proposed in this mighty work that the laws which operate on Earth also apply to the whole Universe. It was a master stroke: and many believe *Principia* to be the greatest scientific treatise of all time.

How could any scientist follow Newton? None tried: there was enough work to do in the century after his death in following through what he had begun. Some astronomers made precise measurements of the positions of bodies to see how they behaved under gravity; this kind of research led to the discovery of the planet Neptune in 1846. Others concentrated on mathematics and theory, establishing the discipline of celestial mechanics. A few worked on the explanation of gravity itself, taking a course which led to Einstein's General Theory of Relativity.

Newton's optical researches inspired larger and better telescopes, now the principles were fully understood. Astronomers such as William Herschel built huge reflectors through which the very structure of our Galaxy could be discerned.

From *De Revolutionibus* to *Principia* is a span of only one hundred and fifty years. By the end of that period, Newton had synthesised the work of his predecessors into a more magnificent vision than even Aristotle's: and once again, astronomy was a science looking forward to the future.

## The Coming of the Camera

By 1840, astronomy was in a buoyant and healthy state. The advances begun by Newton and sustained by his successors meant that the Solar System had been well-charted, and immense star catalogues had been drawn up, noting both star positions and apparent brightnesses. Astronomers had even started to extend into the third dimension, for two star distances had recently been measured. But what were the stars? Could astronomers ever find out? The French philosopher Auguste Comte obviously thought not: for in 1835 he made a pronouncement to the effect that science was quite incapable of giving certain information, and as an example he cited the constitution of the stars.

Two advances in 19th-century science were to prove Comte devastatingly wrong. One – photography – was in its development stages as Comte made his declaration, and by 1839 the first examples had appeared. At last there was a way of permanently recording light-emitting objects, and astronomers were exceedingly quick to take advantage. It must be remembered that all observations up until then had been made visually, by cold, tired and frequently error-prone astronomers. Despite the messy and uncertain nature of early photographic processing, John Draper (in the USA) succeeded in photographing the Moon in 1840, but it was not for a further 10 years that the first picture of a star (Vega) was taken. The early pioneers battled away, however, improving both their chemistry and their technique, so that by the end of the century

astrophotography had become relatively commonplace and stars far fainter than were visible to the naked eye could be recorded.

In the meantime, the desire to build improved optical instruments had led Joseph Fraunhofer in Germany to test prisms of superior glass with a view to making colour-free (achromatic) lenses. Using sunlight as a source, he was surprised to find dozens of dark, straight lines crossing the familiar rainbow spread of colours. He realized at once that these did not arise from lens defects, but were something intrinsic to sunlight itself. By 1817, he had mapped the positions of hundreds of these lines, naming the brightest after letters of the alphabet.

Then, at the University of Heidelberg, the physics professor, Gustav Kirchoff, teamed up with a chemist, Robert Bunsen, to investigate chemical reactions which absorbed or produced light. Kirchoff, a great follower of Newton, suggested spreading out light into its individual colours by means of a prism, and from this, in 1859, the two developed the first spectroscope, in which they passed light first through a narrow slit before it was dispersed. The power in this method lay in its far greater precision: the scientists were able to measure accurately the positions in which the sharp images of the slit fell after dispersion.

Before going further, it is worth explaining why light should be spread out by a prism. Light can be thought of as a series of waves (although, as we shall see later, modern physicists also regard

*Top:* Gustav Kirchhoff, the Professor of Physics at Heidelberg University, explained the dark absorption lines in stellar spectra. He stated that all lines indicated the absence of light at particular wavelengths, indicating the presence of particular chemicals between the source and the observer.

*Above:* Robert Bunsen, one of Kirchhoff's colleagues, is remembered as the inventor of the burner, named after him. It admits air as well as gas to produce a hot, non-luminous flame, which proved useful in spectral research (in fact he did not invent the burner, but popularized its use).

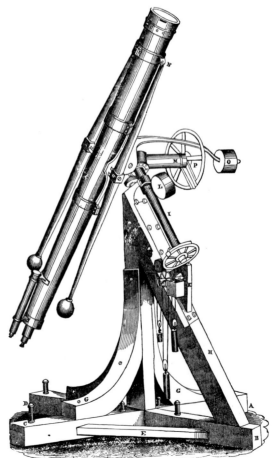

*Left:* Though Fraunhofer died when he was only 39, he was renowned for his instruments as well as for his research. This is a small, 231 mm. refracting telescope he built for the Dorpat Observatory one year before his death.

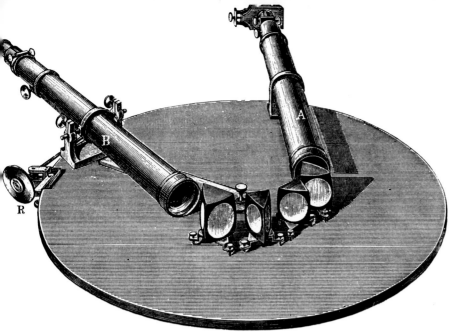

*Top:* Kirchhoff used this instrument to research the solar spectrum. Light enters at B, is focused, and refracted through a series of prisms to the observer's instrument (A), which has a moveable slit at the eyepiece for scanning the spectrum.

*Below:* The spectrum breaks up when passed through a prism and the various wavelengths are bent by different amounts.

*Bottom:* In this early photograph of a spectrum by William Huggins, a nebula (in the centre) is compared with the spectrum of incandescent iron above and below. Four chemicals in the nebula can be identified from their spectral lines.

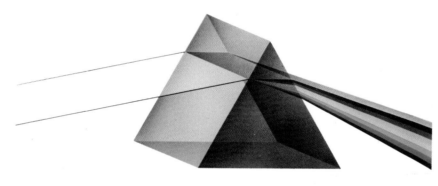

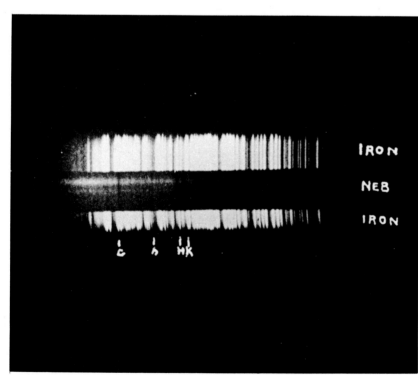

IRON

NEB

IRON

it as a stream of particles) and ordinary 'white' light comprises waves whose wavelengths – the distance from the peak or trough of one wave to the next – range from some 400 nm. (nanometres: thousand millionths of a metre) to 700 nm. Our eyes perceive the shortest wavelengths as blue light, and the longest as red. In everyday situations, all these colours blend together to give white light. But if we pass light through a transparent substance – a lens, a prism or even water – it is bent, or refracted, and short wavelengths are affected more than long ones. The rainbow spread of colours we see when sunlight passes through a prism (whether it is rain or glass) is called a spectrum, and it tells the scientist which particular wavelengths of light are present.

Of course, many other objects besides the Sun also give out light, but not usually in a continuous rainbow of colours. Hot gases, in particular, give out light at certain sharply-defined wavelengths – this is the reason why sodium vapour appears yellow, while mercury looks blue – and a spectroscopist can instantly identify a gas from its characteristic pattern of 'emission lines'. Kirchoff and Bunsen were the first to investigate these spectroscopic 'fingerprints', using Bunsen's newly-invented burner which heated gases to incandescence without producing much extraneous light.

Kirchoff noticed that the wavelength of the sodium line he had measured in the laboratory coincided exactly with the position of the dark line in the Sun's spectrum which Fraunhofer had labelled 'D'. It meant that sunlight must pass through some sodium on its way to Earth: what better place for it to originate than in the atmosphere of the Sun itself? Kirchoff and subsequent spectroscopists went on to find dozens of elements present in the Sun; and one element, helium, was actually discovered in the Sun's outer atmosphere before it was isolated on Earth.

Spectroscopy was the key that astronomers needed to unlock the secrets of the stars and prove Comte wrong. It was not that easy at first, though, for even the brightest star is 10,000 million times fainter than the Sun and its spectral lines are correspondingly dim. But astronomers like William Huggins, working in his own private observatory near London, persevered, and showed that other stars had compositions similar to that of the Sun. Father Pietro Secchi, in Italy, visually observed the spectra of some 4,000 stars – a mammoth task indeed – in an attempt to classify them. And as well as revealing the nature of the stars, spectroscopy turned up an unexpected bonus. The motion of a star directly towards or away from us produces a tiny shift in the position of its spectral lines because of the Doppler Effect (see Chapter 4). This enabled Huggins to derive the first radial velocity of a star – Sirius – in 1868, establishing the basis upon which distances to the most remote galaxies are now measured.

After the 1870s, the rise in the new discipline of astrophysics was rapid. Photographic emulsions became gelatin-based and far easier to use, and so astronomers were able to record spectra photographically and reach much fainter limits. But the greatest boost to the understanding of the nature of stars was in the development of the objective prism – a large, shallow-angled sheet of glass which fitted over the full aperture of a telescope and enabled coarse spectra of many hundreds of stars to be obtained at the same time. In this way, astronomers at the Harvard College Observatory published the details of almost a quarter of a million stars in the Henry Draper Catalogue (1918–1924), which they grouped into spectral classes on the now-standard Harvard Classification System.

By the start of the 20th century, astronomers could measure the composition, temperature, surface gravity, luminosity and velocity of any number of stars. With such a vast amount of data to hand, they were at last in a position to work out what made stars shine, and how they were born, lived and died. From that moment on, astronomy and physics became inextricably intertwined. Astronomers needed to know more about the behaviour of gases in the laboratory before extrapolating to the Universe; while physicists wanted to plumb the near-perfect vacuum of space. These links have been forged even closer as time has passed. The modern astronomer is a true physicist in every sense of the word – and his laboratory is the entire Universe.

To gain more information still, astronomers needed more light. At the beginning of this century, this need led to the building of successively larger telescopes which could reach to fainter and fainter limits. Because a mirror is more efficient at collecting light than a lens, all these great instruments had their origins in Newton's tiny reflecting telescope, adapted to the needs of 20th-century observers. Situated high on mountains away from the glare and pollution of cities, these great eyes on the Universe were, and are, giant cameras.

First of the modern giants was the 1·5 m (60 inch) reflector on Mount Wilson, California – a telescope with a mirror 1·5 m (60 inch) across – followed in 1917 by a 2·5 m (100 inch). Both telescopes were inaugurated by the eminent astrophysicist George Ellery Hale (who founded observatories at Yerkes, Mount Wilson and Palomar Mountain), and began the American specialization in extremely distant, faint objects. Largest of all the American reflectors is the 5 m (200 inch) belonging to the Hale Observatories on Palomar Mountain, and only recently (1976) exceeded in size by the Russian 6 m (236 inch) reflector at Zelenchukskaya.

Strangely enough, the latest generation of telescopes do not follow in this Cyclopean tradition. They have moderate-sized mirrors of around 3.5 m. (140 inch) aperture, and are designed specifically to be used with the widest range of recording and measuring equipment. As we see later, the mirror does only part of the job in collecting the light from a faint object: the astronomer's task today is to record and analyse

it as efficiently as possible. Earlier astronomers had only one means of recording – the photographic plate – but there is now a veritable battery of hardware which can be attached to a telescope. These detectors are many times more efficient than the photographic plate, and so the light-grasp problem is no longer so acute. The new telescopes are a compromise between information acquisition and economics.

Several of the new observatories have found homes in the Southern Hemisphere – Chile has the European Southern Observatory at La Silla and a four metre reflector at Cerro Tololo, while the 3.9 m. (153 inch) Anglo-Australian Telescope is based at Siding Spring in Australia. Contemporary astronomy has trodden the route of the past in this respect, with the northern skies being plumbed first, but these new telescopes will more than redress the balance. As this is being written, the first comprehensive photographic survey of the southern heavens has just been completed, using the one metre Schmidt telescope at Siding Spring. Like its twin at the Hale Observatories, which surveyed the northern skies a quarter of a century ago, this telescope works only as a fast camera, giving undistorted photographs of very large areas of sky. Its result will keep astronomers busy for several decades.

Advances in computer technology have opened the way for a variety of novel telescope designs that have allowed a move away from the conventional single large mirror. The Mount Hopkins Multiple-Mirror Telescope (MMT), in Arizona, is a good example of such a design that has now been in use since 1979. The MMT uses six 1·8 m mirrors mounted in a hexagonal array around a smaller (0·76 m) mirror. Effectively the MMT is six medium-sized telescopes acting together to form the equivalent of a 4·47 m instrument, the light from each of the 1·8 m telescopes being combined at a single focus. The smaller, central telescope is used primarily as a guide telescope for the whole instrument.

Another design uses a segmented reflecting surface – that is a surface consisting of 36 hexagonal elements. The 10 m Keck Telescope being erected at the summit observatory site on Mauna Kea, Hawaii, will be of this form and have the largest collecting area in a single mounting of any telescope to date.

A different type of solution is that adopted for the European Southern Observatory's Very Large Telescope (VLT), where four, identical, single-element 8 m (315 inch) telescopes are specifically designed so that the optical paths may be combined, or used independently if required. All these designs incorporate "active optics", which are essential in such large telescopes. Not only are the mirrors thin, but they are subject to alterations in shape caused by the changes in gravitational loading as the telescopes track objects across the sky. The profiles of the optical surfaces are continuously monitored and adjusted under computer control, so maintaining high optical accuracy throughout observation runs.

**Newtonian**

**Cassegrain**

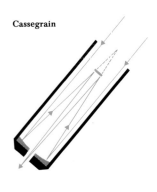

**Coudé**

**Schmidt**

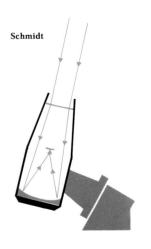

In a Newtonian telescope light is bounced from a parabolic mirror to a focal point near the top of the tube, where a small flat mirror deflects it out through a hole in the tube's side.

In a Cassegrain telescope light is collected on a parabolic mirror at the bottom of the tube. The light is then reflected to a smaller curved mirror near the top of the tube, where it is aimed back through a hole in the centre of the big bottom mirror and brought to focus at the point below the telescope itself.

The coudé focus is like the Cassegrain except that it has a flat mirror to deflect the light sideways. This mirror moves, compensating for telescope motion so that the resulting light beam can be kept in a fixed position.

The Schmidt telescope combines features of both reflector and refractor instruments. Used solely for photographic work, the Schmidt first passes light through a lens which corrects spherical aberration – the loss of sharpness produced by the different focal lengths of different parts of the same lens then bounces the light off a large mirror to a point of focus on a curved photographic plate.

The 48-inch (1.22 m.) Schmidt telescope at Siding Spring, New South Wales, Australia, was installed at the Anglo-Australian Observatory in 1973. It is carrying out a survey of the southern skies to match that of the northern hemisphere made by the Mount Palomar Schmidt of the same size in the early 1950's.

A worker, standing by the open framework of the Mount Palomar 200-inch (508 cm.) reflector, shows the scale of the instrument. The mount behind him is to hold the mirror which will be set at the end of the 110-ton tube. The telescope, which opened in 1948, is so delicately balanced that one hand could turn it on its oil-pad bearings.

A radio-telescope is a huge antenna for collecting weak radio signals from space. This type uses a large dish (Jodrell Bank is an example) to concentrate the waves on the antenna proper, mounted above the centre of the dish. The signals are first amplified by a receiver, then sent to a simple computer which sorts out static. A recorder finally transcribes the signal on a graph.

Karl Jansky, an American communications engineer and founder of radio-astronomy, pushes round the radio receiver, which, in the early 1930's led him to be the first to identify a radio source beyond the atmosphere.

On the whole, most modern telescopes bear quite striking resemblances to one another, as visitors to observatories will have noticed. This is partly related to the limited number of telescope manufacturers and designers, but has even more to do with the nature of a telescope's work. Basically, the instrument must collect light from a faint source and focus it on to a detector. However, its job is complicated by the Earth's rotation, and the telescope has to accurately track the object under surveillance as it appears to move across the sky. Most telescopes manage this by being equatorially mounted: part of the mounting is precisely aligned with the celestial pole and the instrument pivots around this 'polar axis' to follow the stars. As objects often need to be monitored by equipment for periods of several hours, the telescope needs to be very smoothly and precisely driven, without any mechanical flexures or vibration. Until recently, this was always a taxing engineering feat, but the widespread application of computers to astronomy has meant that it is becoming a problem of the past. In both the Anglo-Australian Telescope and the Russian six metre reflector, a computer takes care of the tracking, correcting continuously for changes in the angle of the telescope, positioning of equipment, atmospheric refraction and many other factors.

For all its complexity and expense, a modern telescope's main function is simply to gather light. At its focus there is a greatly brightened image of some distant celestial object, and the astronomer must decide how to analyse it. Traditionally he viewed it through an eyepiece, to obtain a highly magnified image; but the human eye is not a very sensitive detector – it cannot store up the light falling on it – and the analysis of the image cannot help but be subjective. Moreover, it is difficult to measure positions accurately at the telescope.

Photographic plates are now universally used to record star images in permanent and easily measured form. Modern photographic emulsions are 50,000 times faster than those of the last century, latest techniques including the baking of plates under dry nitrogen for several hours before use to increase their sensitivity. Cooling them with liquid air during the exposure also helps when faint objects are under study, for it alleviates 'reciprocity failure', the unfortunate tendency of photographic emulsions to need disproportionately longer exposure times at low light levels.

Even the best photographic emulsions are fairly inefficient light detectors, however. To get some measure of efficiency, astronomers utilize the fact that light is not just a wave-motion, but these waves come in separate packets of energy, called *photons*. Normal astronomical plates record only one in a hundred of the photons which fall on them: 99 per cent of the light goes to waste.

In recent decades, astronomers have been experimenting with electronic devices which will alleviate this chronic wastage, and this has led to the trend away from large telescopes: a perfect

detector of light mounted on a 50 cm. (20 inch) telescope would be as sensitive as the 200 inch telescope equipped with standard photographic plates. The first highly successful device was the photomultiplier tube, still an indispensible tool, which can detect one photon in six. Photomultipliers complement, rather than replace, photographs because they cannot form an image: they simply measure (very accurately) the brightness of stars or galaxies. Astronomers use these devices to maximum advantage by confining their measurements to one colour at a time, cutting out the others with a filter.

Since photomultipliers can only measure star brightnesses, the past decade has seen a sustained effort in adapting photo-electric brightness-measuring devices – photometers – to record two-dimensional images. This drive has been not so much to obtain more accurate pictures of extended objects like galaxies and nebulae, but more to record spectra precisely.

Astronomical spectrographs no longer use glass prisms, which absorb valuable light. The wavelengths are separated by reflection off a *diffraction grating*, a very finely ruled sheet of glass. Due to the wave-nature of light, each wavelength can only reflect off in a few specific directions, so the light is split into several spectra, lying in different directions from the grating. By shaping the fine grooves suitably ('blazing') astronomers can ensure that almost all the light falls into just one of these spectra. The task is then to record this spectrum in the most efficient and accurate way.

Although most spectrographs still use photographic plates, electronic devices are very much on the way in. One of the most successful is the Spectracon, in which a cathode emits electrons independently according to the light falling on each part of it. The electrons are then focused to form an exact image of the astronomical object or its spectrum.

Since the Spectracon tube is evacuated, while the photographic film must be outside – for technical reasons as well as ease of handling – the tube ends in small mica window, against which the film is pressed. The window must be very thin so it absorbs few electrons, but this makes it weak – the untimely demise of a Spectracon is usually caused by someone accidently puncturing the instrument's window and so letting air into the tube.

A highly sophisticated system – arguably the best in the world – is the University College London's Image Photon Counting System. Here the screen of the initial image tube is scanned by a television camera, and a computer automatically measures each spot of light on the screen to ensure it is actually caused by an electron from the photocathode, and not by a stray electron or charged atom in the tube.

A development of great significance to modern astronomy was the introduction of super-sensitive, and highly efficient, electronic detectors called charge-coupled devices (CCDs). These are two-

dimensional array detectors which use silicon semiconductor technology to provide astronomers with a way of gaining not only more detailed images of the objects they are studying, but also enable them to manipulate the data to extract much detail. The collecting surface of the CCD is divided into an array of picture elements, or pixels, 250,000 or more in a single device. The incident image is focused on the surface of the CCD from where it can subsequently be electronically processed for display on a monitor screen, converted into a photographic print or stored for later analysis.

The need for clear skies has driven astronomers to seek observatory sites in remote areas of the world. But always, there has been the Earth's atmosphere to contend with. Two sites in particular deserve special mention. Both are located at high altitude and the results coming from the various telescopes at the observatories attest to the wise choice of the sites as the major centres of international astronomical research. One of the sites is on the 4200 m (15,000 feet) summit of the extinct volcano, Mauna Kea in Hawaii, where there is a growing number of telescopes that have been designed to use the latest technology to gain the maximum amount of

*Top:* The Arecibo radio-telescope in Puerto Rico is the world's largest radio-astronomy dish, measuring 1,000 feet (305 m.) across. It is slung hammock-like in a natural dip. The receiver, mounted 870 feet above the wire-mesh surface and held in position by the side trusses, can be moved to give a 50° north–south coverage.

*Above:* These five dishes are part of the Mullard Radio Observatory, Cambridge. They, and others like them, are used to provide very long base lines to obtain accurate fixes on distant objects.

The Pic du Midi Observatory,
set at 9,390 feet (2,862 m.) in
the French Pyrenees, shows the
extremes to which astronomers
must go if they are to avoid the
turbulence and pollution of the
lower atmosphere. To use the
new 78-inch (200 cm.) reflector,
astronomers have to brave
arctic conditions (*inset*).

information. The skies above Mauna Kea are
excellent for infrared and millimetre wave
observations, due to the lack of water vapour in the
atmosphere at that altitude. The other major,
growing mountain top observatory is on La Palma,
in the Canary Islands, off the west coast of Africa. A
number of countries are negotiating for permission
to construct instruments at the site in view of the
clarity of the sky. The 2·4 m (98 inch) Isaac
Newton Telescope was moved there, from its
original home at Herstmonceux, Sussex, in
England, in the early 1980s. The United Kingdom
has now also constructed the 4·2 m (160 inch)
William Herschel Telescope (see page 170) on La
Palma. The early results from the new instrument
are very promising and there is great hope for the
future.

## Whispers from Space

Our preoccupation with the ingenious devices
used by modern optical astronomers has tended
to obscure the fact that light gives us only one set
of information. Glancing through this encyclo-
pedia, the reader will notice that objects in
space emit all sorts of other radiations as well –
radio waves, X-rays and infrared radiation to
name but a few. Light is just one very narrow
band of wavelengths belonging to the whole
electromagnetic spectrum – and the recent
revolution in astronomy has been mainly centred
on exciting discoveries made at other wavelengths. The entire electromagnetic spectrum
ranges from the ultra-short gamma-rays, whose
wavelengths are measured in terms of a million-
millionth of a metre, up to radio waves hundreds
of metres long. (There are no actual physical

To analyse the light that they
gather in their instruments,
astronomers are increasingly
turning to electronic data
processing and computer
techniques. A prime example is
the Image Photon Counting
System (*far right*), developed by
Alec Boksenberg at University
College, London, which records
individual photons. The picture
at *right*, showing photon
'impacts', represents less than
1 per cent of the IPCS's field.
With it, any light source can be
surveyed at any frequency over
any time period – a vast
advance over photographic
methods.

boundaries between these radiations, apart from their range of wavelengths or frequency. The various names have arisen because the waves are generated by different processes, and we need different kinds of detectors to receive them.)

Although light waves cover a band of the electromagnetic spectrum which amounts to only a thousand-million-million-millionth ($10^{-21}$) of its total spread, they assume such importance to us because they are one of the few radiations to penetrate our atmosphere. Shorter-wavelength ultraviolet rays (with wavelengths less than 300 nm.) are blocked by the atmospheric ozone layer, whereas the longer infrared radiation (wavelengths greater than 1,000 nm.) is mainly absorbed by water vapour. The atmosphere is once again transparent to radiation between 1 cm. and 30 m., the so-called 'radio window'. Longer wavelength radio waves are prevented from reaching the ground because they bounce off the upper layers of the ionosphere – a belt of charged particles at the top of the atmosphere whose reflecting properties (from the lower layers this time) make worldwide broadcasting possible.

Because of the atmosphere's transparency to radio waves, it is not surprising that they were the first of what were once called the 'invisible radiations' to be exploited by the astronomer. Yet their discovery came about by accident in 1931, when Karl Jansky of the Bell Telephone Laboratories, looking for static from thunderstorms, detected a faint background of radio 'noise' which rose and set with the stars. But it was not generally until after World War II when radar – the bouncing of radio waves off objects to determine their position – was developed that sufficient practical knowledge existed to support radio astronomy.

Modern radio telescopes work essentially like giant radio sets, the only difference being that the radio waves they receive come from natural transmitters in space. But a radio telescope needs to be millions of times more sensitive than even the most sophisticated hi-fi receiver – the total amount of energy picked up by all the radio telescopes in the world, over the entire history

of radio astronomy, is less than the energy you would expend in lifting up a feather.

Much of the ingenuity in this field of astronomy lies in receiving and amplifying these whispers from space. Most radio telescopes work rather like huge optical reflecting telescopes, collecting the radio waves in a curved dish and focusing them on to an aerial in the centre. From the aerial, the waves are fed to a receiver, which often needs to be designed with the characteristics of particular types of radio source in mind. Every attempt is made to cut down spurious 'noise' at this stage, by methods ranging from cooling the receiver with liquid air to incorporating a maser (the microwave equivalent of a laser) into the circuitry. After being amplified, the signals are generally recorded directly on magnetic tape and analysed by computer, which can process the information at a rate far surpassing any human researcher. Such automation has always been the norm in radio astronomy: and its fearless exploitation of computers and modern electronic techniques has had an invigorating effect on optical astronomy, as we saw earlier.

In some ways, though, the radio astronomers' lot is not a happy one, compared to that of the optical astronomer. The latter builds large telescopes not only to capture more light, but also to see finer details. Eventually, however, the optical astronomer gets no return for a further increase in size, because very small images become blurred out by irregularities in the atmosphere. However, the resolution also depends on the wavelength, and so radio astronomers, using waves a million times longer than light, need to build radio telescopes to an appropriate scale if they are to see the same degree of detail as an optical astronomer.

Although this low resolution was a great problem in the early days, radio astronomers have become adept at getting round it. The most elegant solution was pioneered by Britain's Astronomer Royal, Sir Martin Ryle, who electronically linked the output from a line of small radio telescopes, and allowed the Earth's rotation to swing them round as they tracked a source. By stacking the results in a computer, and altering the spacing of the dishes each day, Ryle was able to achieve the resolving power of a single radio dish with a diameter equal to the length of the array. This technique of Aperture Synthesis allows Ryle's Five Kilometre Telescope in Cambridge to resolve detail as fine as would be seen in a good optical telescope. And by linking up radio telescopes thousands of miles apart – perhaps on different sides of the world – radio astronomers to pin down extremely small-scale structures even if they cannot get a proper picture. This technique, called Very Long Baseline Interferometry, is telling astronomers much about the sizes of emitting regions in quasars and radio galaxies, and even about the centre of our own Galaxy.

From remote galaxies we return to the Solar System, and to an illustration of astronomy in an

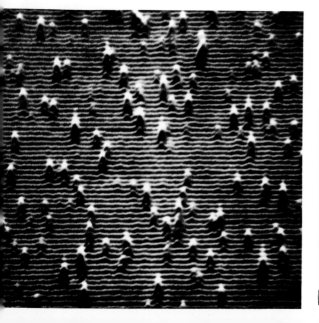

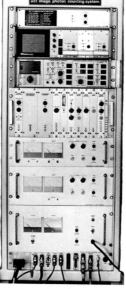

active, rather than a passive role. The technique of radar is now being used to probe the planets, and has already met with considerable success. Early work was concerned with measuring planetary distances – a relatively simple matter of 'timing the echo' – but soon extended to timing the spinning of planets from the Doppler Effect this causes in the reflected radio pulse. In this way, the rotation periods of Venus and Mercury were discovered (in 1962 and 1965 respectively). Now radar is a valuable tool in planetary mapping, particularly in the case of cloud-covered Venus, where radar echoes have located craters and mountain ranges. Spacecraft orbiting Venus have also used radar to map the surface features with great effect.

The region between light and radio waves in the electromagnetic spectrum – from 0.001 to 1 mm. – is the domain of infrared radiation. This

is familiar to us all as 'heat', and there are several types of natural infrared sources in space, such as the clouds which surround forming stars. The infrared astronomer is beset by the problem that, although some radiation does get through our atmosphere, most is absorbed by water vapour. So infrared observatories are situated on bitterly cold mountain tops where as much water vapour as possible is frozen out. However, the astronomers who man the Kuiper Airborne Observatory – which flies above most of the Earth's atmosphere – have a more comfortable time. An additional problem to be faced by the infrared astronomer is that he himself, his equipment and the observatory all emit relatively vast amounts of heat, creating brighter infrared sources than any thing in the heavens. He is rather in the position of an optical astronomer attempting to observe a faint star under floodlights. All

The Roque de los Muchachos Observatory, La Palma. From left to right can be seen: the dome of the William Herschel Telescope, the domes of the Isaac Newton Telescope and Joseph Kaptyen Telescope and the Swedish Solar Tower. *Inset*: The 2.5 m Isaac Newton Telescope inside its dome at Roque de los Muchachos Observatory.

he can do in the circumstances is to resort to a number of sophisticated electronic techniques, and use cooled detectors especially sensitive to infrared radiation.

All wavelengths shorter than light – ultra-violet radiation, X-rays and gamma-rays – are absorbed by the atmosphere, and so the analysis of these radiations has had to await the advent of the Space Age. In many respects, these branches of astronomy share a similar approach. Their detectors need to be of a compact and robust design to be sent into space; and they may well have to vie for accommodation aboard a satellite which, on economic grounds, may be crowded with more than a dozen competing experiments.

First of the trio was X-ray astronomy, which began in 1962 with the accidental detection from a rocket of an X-ray emitting source in Scorpius (Sco X-1). Research groups in the UK and USA

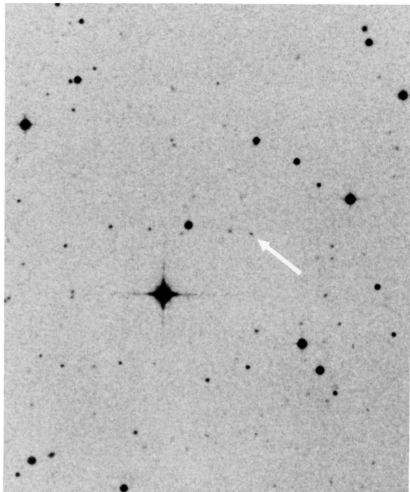

flew many more rockets with X-ray detectors on board to learn more about these sources, but until 1970, when the first X-ray satellite (UHURU) was launched, the total flight time only amounted to minutes. UHURU, and the British satellite Ariel V, launched four years later, have made comprehensive surveys and have detected many new – and often transient – X-ray sources.

What of the future? It seems certain that the destiny of astronomy lies in space; not simply to detect radiations otherwise unattainable, but to branch out in a hundred other directions. Who can tell what revolutions the Space Telescope – the first optical telescope scheduled to soar above Earth's grey, churning atmosphere – will bring when it is launched early in the 1980s? And then there is 'active' astronomy too; the physical exploration of planets by probes, and soon, Man. Next will come the stars.

Our story has taken us through 5,000 years. We have traced the course of an objective science, and encountered superstition, fear, mistrust and inflexibility on one hand; and the glorious rewards of intellectual achievement on the other. And now we are at another watershed in astronomy. For thousands of years it has been a purely Earth-based science. But space is beckoning: a new era has begun.

In their quest for knowledge, astronomers have been pushing back the horizon of the visible universe. This quasar (see Chapter 7), with the unglamorous name Q0051-279, is the most distant object in the universe so far detected. It has a redshift of 4.43, which places it about 13,000 million light-years from us, and it is receding at 93 per cent of the speed of light. The light from this quasar set out on its journey when the universe was just one-tenth of its present age.

# 10/PROBES AND SATELLITES

The launching of the first Earth satellite by the Soviet Union in 1957 was one of the greatest shocks sustained by the West since the war. The US responded with a massive investment in rocketry and assorted space hardware. Beginning with the Redstone launcher – now enshrined in a "historic Redstone test site" (left) – the most dramatic manifestation of this investment was the lunar landing in 1969. However, the developments in unmanned spacecraft – in particular Russia's Lunar rover and the US's Mars Lander – have shown that Man can explore the Solar System indirectly, with a corresponding saving of money and, almost certainly, lives.

**D**ue to limitations imposed by the Earth's atmosphere, ground-based astronomers can only make studies of astronomical objects in a relatively small number of narrow wavelength bands in the spectrum. As we saw in the previous chapter, telescopes can be sited at observatories high up on mountain tops, carried to high altitudes slung underneath helium-filled balloons or in high-flying aircraft, such as the Kuiper Airborne Observatory. However, even with all these methods, there is still some atmosphere above the instruments to contend with.

The next, most obvious solution for overcoming atmospheric problems is to place astronomical instruments high above the "sensible" atmosphere aboard Earth-orbiting satellites. This has been done, with great effect, on numerous occasions over the past twenty-five years, thus allowing astronomers to explore the entire spectrum, from low energy radio waves to high energy gamma-rays. Observations across the entire spectrum have revealed new aspects of what were thought to be

ICE encounter with Comet Giacobini–Zinner on September 11, 1985. The spacecraft passed through the comet's tail, 10,000 km (6200 miles) from the nucleus.

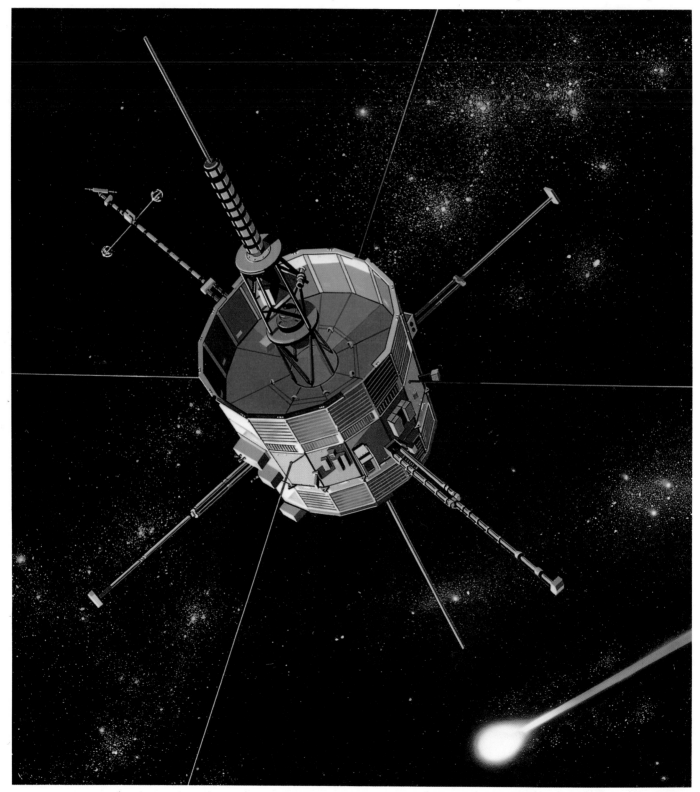

well understood phenomena, and have even revealed objects that we could never even have discovered if it were not for the advent of the astronomical satellite.

Astronomers are not restricted to gathering data from just the Earth's orbit today. Spaceprobes have been sent out into the depths of the solar system to explore the interplanetary environment, as well as to investigate the nature of the planets and comets we have known for so long, but whose details, in hindsight, we knew hardly at all. So far, six planets and two comets have had unmanned spaceprobes visit them, radioing back to Earth-based astronomers the much sought-after images and other vital data that will enable us to better understand the solar system.

It is not possible here to cover all the spaceprobes and astronomy satellites that have been launched, and the results they have returned, but we shall be looking at some of the more prominent ones.

## Missions to the comets

The first ever spacecraft to encounter a comet was not, in fact, designed for the purpose. The International Cometary Explorer (ICE) space-craft was originally designed and equipped to study magnetic fields, plasma, electric and magnetic waves, atomic particles and radiation. Its original mission did not require any imaging capability, so no cameras were aboard.

In 1978, the spacecraft, then called ISEE-3 (International Sun–Earth Explorer 3), was laun-ched to orbit a point in space located 1,500,000 km (930,000 miles) from the Earth on the line extending towards the Sun. At this libration point, the net force on the spacecraft, including the gravitational attractions of the Sun and Earth is zero. ISEE-3's job was to monitor the solar wind approaching the Earth, while its sister craft, ISEE-1 and 2, explored its effect on the Earth's radiation belts and other parts of the magnetosphere.

When its first mission was completed, ISEE-3 was moved to investigate the Earth's magnetic tail, which stretches for more than 250 Earth radii in the direction away from the Sun. The new orbit of ISEE-3 allowed it to make detailed measurements of the "tail" region at points between 80 and 220 Earth radii along its length.

Mission scientists realized that ISEE-3 would be the almost ideal spacecraft to send for a rendezvous with a comet and return data on the environment within the cometary tail. So, following a complex series of orbital manoeuvres, involving no less than five lunar fly-bys, ISEE-3 swung to within 120 kilometres (75 miles) of the Moon's surface, on 22 December 1983, in order to gain a gravitational boost which would help it on its way. As it departed the Earth–Moon system, ISEE-3 received its new name to better reflect its new mission.

The target selected for ICE was the small, short period comet, named Giacobini–Zinner. During the approach to the encounter on 11 September 1985, ground based optical astronomers tracked the comet and were able to supply information that helped spacecraft controllers steer ICE on the

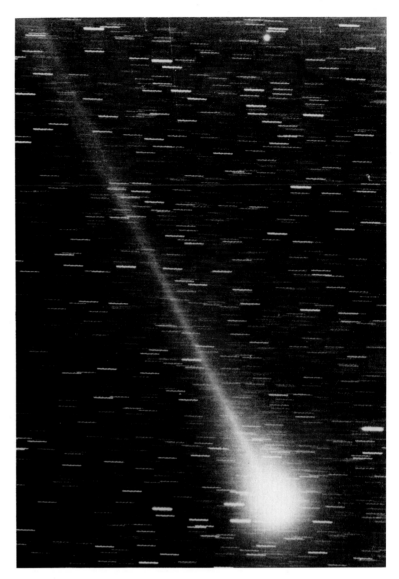

Comet Giacobini–Zinner on October 21, 1959.

precise trajectory needed for the tail interception.

The encounter went almost exactly as planned, and the results were a surprise to almost all cometary scientists. There was far more dust in the tail than had been predicted. ICE detected traces of carbon monoxide and various types of water molecule, both of which were particularly abun-dant in the middle region of the tail. This was strong evidence for the theory that the nucleus of a comet is made up of frozen water mixed with a great deal of dust and rocky material left over from the formation of the solar system.

About six months later, probably the most famous of all comets, Halley's Comet, was also visited. Not by one spacecraft, but by five – one from the European Space Agency (ESA), two from the Soviet Union and two from Japan. Not all five were to actually encounter Halley, but all were to pass close enough to reveal its secrets to varying degrees. On 6 March 1986, the first of the Soviet probes, Vega 1, flew past at a distance of just 10,000 km (6200 miles) from the nucleus. Both Japanese craft, Sakigake and Suisei, made their long distance passes on 8 March at distances of 4 million kilometres (2.5 million miles) and 200,000 km (124,000 miles) respectively. On the next day

Giotto prepares for its encounter with Halley's Comet after its eight-month voyage through interplanetary space. With a closing speed of 68 km per second, Giotto was not expected to survive the passage through the head of the comet. However, though badly battered, the spacecraft did survive.

Vega 2, the second of the Soviet craft, passed even closer to the nucleus than had its twin. However, by far the most dramatic, and probably the most rewarding of the encounter missions, was that carried out by ESA's craft, Giotto.

At a relative speed of some 68 km (42 miles) per second, Giotto passed to within just 500 km (310 miles) of the nucleus. Fears that Giotto would be totally destroyed by high speed impacts with cometary dust particles came to nothing. After a 700 million kilometre (430 million mile) journey, Giotto had time to transmit several images of the nucleus of Halley, before its camera was damaged and it was knocked about so much that tracking stations on Earth lost contact temporarily with the now famous craft. Brief details of the results of these missions are given in Chapter 4.

### Mars missions

Mariner 4 was the first man-made object to fly past Mars. On 14 July 1965, it passed as close as 9800 km to Mars and sent back 21 pictures of the surface. It would be an understatement to say that the pictures were a surprise to the science teams back on Earth. The surface of the red planet appeared to be covered with craters not too dissimilar to those of the Moon. In 1969 images were sent back by Mariners 6 and 7 from the closer distance of 3540 km (2200 miles). The 201 pictures returned revealed not much different information to those from Mariner 4.

What was needed was an orbiting spacecraft so that more of the surface could be photographed and in greater detail than had been possible with the cruder cameras of the early missions. So, on 14 November 1970, Mariner 9 became the first man-made craft to enter orbit about another planet in the solar system. As soon as the two TV cameras were switched on, nothing could be seen because a planet-wide dust storm was occurring that all but obscured the surface. Such storms had been known to occur, but sheer bad luck meant that one was in progress as Mariner 9 arrived. While it appeared, from the earlier Mariner pictures, that Mars was a dead world in terms of the possibility of any form of life, and perhaps geological activity, it certainly was active meteorologically!

As the dust settled, it became obvious that the earlier Mariner craft had simply not shown enough detail. One of the earliest surprises of the mission came on 7 January 1971. What turned out to be the largest known volcano, Olympus Mons, was discovered. Various measurements showed that this feature was a staggering 24 km (about 15 miles) high.

It had been expected that Mariner 9's mission would last for 90 days and allow up to 70 per cent of Mars to be mapped. However, as so often happens in space science, the equipment and the results were far better than expected, and the mission lasted for 349 days. During this time the whole planet had been photographed, a total of 7329 images. But what of the question of life? Not even Mariner 9's cameras and sensors were capable of even attempting to find any signs.

The search for life would have to be made on the spot. The first attempt to do just that was made only days after Mariner went into orbit, but it took a year to achieve success. After two failures, the Soviet Union landed Mars 3 on the Martian surface on 2 December 1971. Ninety seconds later, the 350 kg (772 lb) probe's camera started transmitting. But the signals lasted only 20 seconds and the picture, to use the official description "did not reveal any noticeable difference in the contrast of details" (it was blank!). Circling above, Mars 3 (like Mars 2 before it) wasted all its film on a planet hidden beneath 8 km (5 miles) of dust because it could not be reprogrammed to wait out a storm. The rest of the Mars series were all failures. Mars 4 missed the planet completely; Mars 5 added little to Mariner 9's findings; the Mars 6 lander crashed and Mars 7 fired its lander into solar orbit. That cosmic convoy ended the Soviet Union's phase of Mars exploration for a while.

But in America space scientists were preparing their own search for life on the Red Planet with the Viking probes. The first two contained astonishing automatic laboratories designed to test the Martian soil for the existence of life.

On 9 June 1976, Viking 1 made the longest deep-space engine burn in history (38 minutes) to place itself in orbit around Mars. The 3399 kg (3·3 ton) Viking was three times heavier than Mariner 9 and took the best part of a year to complete its 708 million km (440 million mile) journey. The first two weeks in orbit were spent searching for a smooth landing site amid the jumble of boulders, mountains and ravines. Controllers found it, with the help of Viking's cameras and Earth-based radar, on the western edge of the "Gold Plain", Chryse Planitia.

Three hours before landing the lander separated from the Viking orbiter and fired its retrorockets for 20 minutes. Slowed first by the thin atmosphere, then by giant parachutes and finally, after dropping its heat shield, by three main

descent engines, the three-legged robot – a 2 m (6 feet) tall, hexagonal aluminium box measuring 3 m (10 feet) across and weighing over 575 kg (half a ton) – settled on the Martian surface at 4.53 a.m. (California time) on 20 July 1976 – the seventh anniversary of the first manned Moon landing. Twenty-five seconds after touchdown, one of two cameras was switched on to record the first black and white picture from the surface of Mars. Six minutes later it began a 300-degree panorama of the Mars-scape.

The pictures were astonishingly good. The later colour pictures were even more remarkable, proving at last that the surface of Mars really is red.

We knew so little about the Martian surface that no one dared rule out the possibility of seeing a landscape teeming with plant life. There were no plants. The next step was to look for microscopic life forms in the soil. To do this Viking had three

Atop a Titan-Centaur rocket, Viking 1, containing the first Mars lander, is blasted off from Kennedy Space Center on August 20, 1975.

The Viking Mars lander (*below*) proved a triumph for the US. Its sophisticated scientific package not only took classic pictures of the Martian landscape (*right*) and sunset (*far right*), but also tested for the existence of life (see Chapter 6). For the sunset shot, the camera scanned from the left as the Sun dipped below the horizon and took ten minutes to complete the 120 degree coverage from left to right; hence the better lit section on the left and the stroboscopic effect in the Sun's glow. Parts of the lander can be seen in the centre and right.

tiny biological laboratories powered, like everything else aboard the lander, by a pair of small nuclear generators. Electricity was at such a premium that engineers designed the laboratories to operate on little more power than a domestic lightbulb. Yet they were designed to detect life of any kind, even in soil many times more barren than Earth's most barren desert.

All three experiments worked on the assumption that Martian life would be similar to organisms on Earth, simply because it was impossible to fit more ambitious experiments into less than 0·28 m³ (cubic foot) inside the lander's body. Soil samples for each experiment were lifted from the surface during Viking's ninth day on

Mars by a shovel at the end of a 3 m (10 feet) extendable boom. The results were dramatic (see Chapter 6) but inconclusive.

The perplexed scientists – and the world – would have to wait and see what happened with Viking 2, when it reached Mars. The second lander touched down on Utopia Planitia (Utopia Plain) on 3 September 1976. It was dumb for several hours before landing with a radio failure aboard the orbiter, but it descended automatically and communications were restored eight hours after touchdown. Viking 2 repeated the companion's search for life, but the results were equally uncertain. Something, animal or mineral, was giving unexpected readings – and scientists

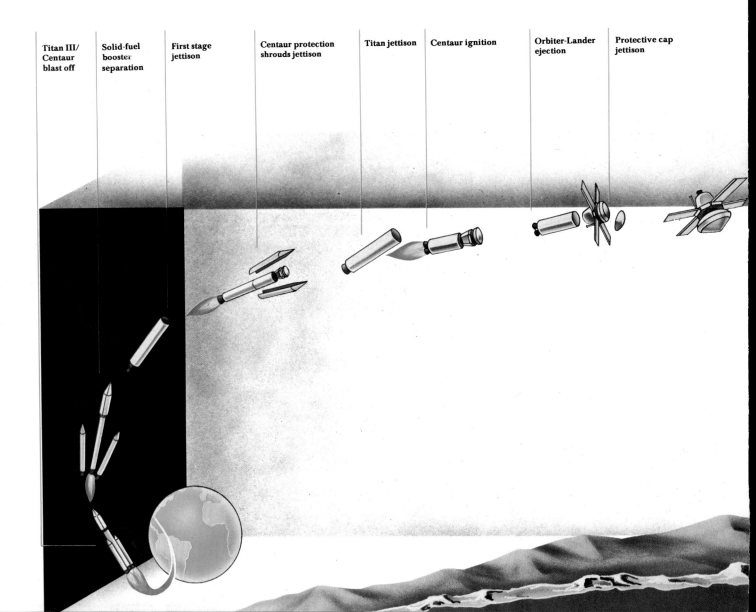

| Titan III/ Centaur blast off | Solid-fuel booster separation | First stage jettison | Centaur protection shrouds jettison | Titan jettison | Centaur ignition | Orbiter-Lander ejection | Protective cap jettison |

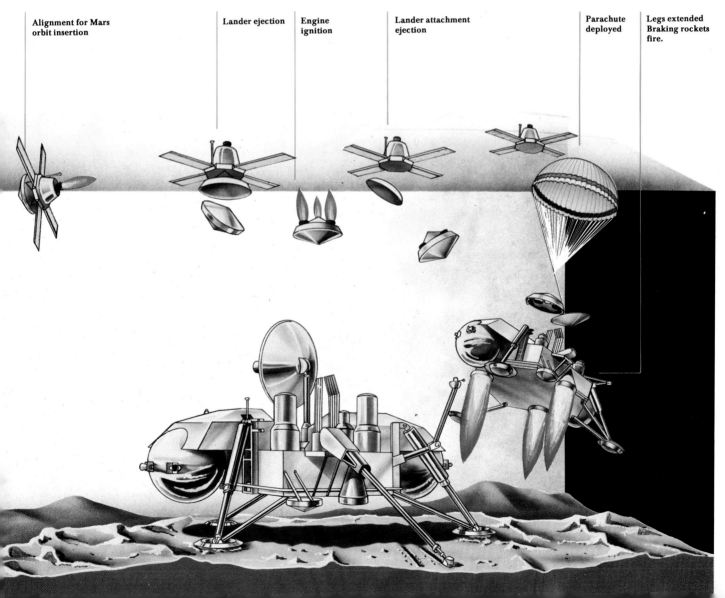

**Alignment for Mars orbit insertion**

**Lander ejection**

**Engine ignition**

**Lander attachment ejection**

**Parachute deployed**

**Legs extended Braking rockets fire.**

concluded sadly that it was probably not a life form. The Viking 2 orbiter was switched off on 25 July 1978, after orbiting the planet 706 times, but the other orbiter and the two landers continued operating well into their third year in space. The Viking 2 lander even sent pictures of ground frost on a chilly (−80°C;−114°F) winter morning. The two landers were shut down in the spring of 1979.

The Viking search for life ended officially on 1 June 1977. They hadn't proved there is life on Mars, but nor had they proved there isn't. Viking's 3 m (10 feet) reach meant that between them the landers could only search 17·7 m² (190·6 square feet) on a planet of many millions of square kilometres.

## Mercury and Venus missions

Mariner 2 had shown Venus to be a most forbidding place, but Soviet scientists decided to take up the challenge of a world where a man would be at once poisoned in a lethal atmosphere, fried in the oven heat and squashed by atmospheric pressure many times that of Earth's.

The early Venus flights were notable technological failures. Between February 1961 and February 1966, nine Venus probes failed abysmally – three flew silently past the planet with broken radios and six never made it out of Earth orbit. On 1 March 1966, Venera 3 crashed onto Venus, the first man made object to reach another planet. But this craft too fell silent before reaching its target, making it a scientific failure.

In October 1967, Venera 4 gave an indication that success was possible by transmitting data from inside the Venusian cloud cover. It penetrated 27 km (17 miles) into the atmosphere and worked for 94 minutes before being crushed by the terrible pressure. (A day later, America's Mariner 5 passed by 4000 km (2500 miles) from the surface.) In 1969, Venera 5 and 6 both penetrated to about 25 km (16 miles) on the light side of the planet before being crushed, and on 15 December 1970, a tougher and more sophisticated Venera 7 transmitted the first signals from the surface of another planet. Venera 8 repeated that success on 22 July 1972.

The next Venus probe was Mariner 10, an American probe which killed two birds with one stone by flying past Mercury too. The first two-

planet explorer passed within 5760 km (3580 miles) of Venus on 5 February 1974, transmitting the first pictures of the planet's lethal cloud cover. Venusian gravity then bent the probe's trajectory for its rendezvous with Mercury, a trick known as a "slingshot manoeuvre".

The innermost planet has always been difficult to study because it is so close to the blinding Sun. But on 29 March 1974, Mariner 10 gave us our first view of the Moon-like surface, from a distance of only 271 km (168 miles). Its orbit was so impeccably calculated that the probe, now in solar orbit, passed Mercury twice more (September 1974 and March 1975), sending back more than 10,000 pictures, together with magnetic and radiation measurements – all for a mission cost of £41·6 million ($100 million).

By this time, Soviet scientists felt certain they had mastered the problems of Venus's hellish environment. On 18 June 1975, they launched a giant Proton rocket from Tyuratam carrying a Venera probe four times heavier than earlier craft – a 5000 kg (5 ton) orbiter and lander called Venera 9, complete with searchlights to counter the pall of darkness that the planet's clouds presumably cast.

On 22 October 1975, the orbiter became the first artificial Venusian satellite while the 1560 kg (1½ ton) lander began a descent unique in the history of lunar or planetary exploration, using a combination of six parachutes and its own shape to pass quickly through the hottest parts of the atmosphere. The hair-raising drop ended at 2.13 p.m. (Moscow time), and 15 minutes later Venera 9 gave man his first view of another world: a landscape of large and small rocks, but without much dust and rocky debris. One big surprise was that Venera's searchlights were unnecessary – Venus was "as bright as a cloudy day in Moscow".

Venera 10 arrived at Venus three days later and touched down 2200 km (1370 miles) away. Venera 9 had landed on a high plateau (possibly the slopes of a volcano) with "young" angular rocks scattered about, while Venera 10's pictures suggested a stony desert, with old weathered rocks like pancakes. The 53 minute lifetime of Venera 9 and 65 minutes life of Venera 10 have revolutionized our understanding of Venus. Man can now work from first-hand knowledge, not speculation.

Venus's rock-strewn surface was recorded for the first time in June 1975 by Russia's Venera 9 soft-lander. Only one other picture (by Venera 10) has been taken of the surface, for the conditions are truly frightful. In heat that will melt lead and atmospheric pressure 90 times that of Earth, for the probe to operate at all, even for a few minutes, was a triumph.

The next Venus mission had a more specialized role: its purpose was to use the "Evening Star" to tell us something about our Earth. American scientists reasoned that a study of Venusian weather patterns would yield useful insights into our own complex weather system. That was the mission of Pioneer Venus 1978, a dual-launch project begun in the summer of 1978. An orbiter began circling Venus in early December while another spacecraft released four atmospheric probes – one large and three small – into the Venusian cloud cover. It was one of the cheapest interplanetary missions ever attempted, because everything had been tested on earlier flights and no prototypes were built.

## The outer Solar System

The outer planets, those beyond Mars, were not to be ignored. Once the navigation and control techniques necessary for flights to the Moon and the planets of the inner Solar System had been learnt, the time had come to apply that expertise to the spacecraft study of the gas giants – Jupiter, Saturn, Uranus and Neptune. Plans were drawn up for a mission that would send a spacecraft to fly past each of these worlds, to send back information about them and their attendant moons. The "Grand Tour", as the project had been dubbed, was abandoned as a result of a lack of Federal funding. However, all was not lost. The highly successful Pioneer series led to two of its number, Pioneer 10 and Pioneer 11, being sent out to Jupiter and then, just a couple of years later, Pioneer 11 became the first man made object to encounter, and fly past Saturn. The images returned surpassed, by far, anything that could be obtained by even the most powerful Earth-based telescopes. The complex weather systems of both planets were revealed in great detail, and the rings of Saturn were found to have a more complicated structure than had been previously thought.

Following hot on the heels of the highly successful Pioneers were Voyagers 1 and 2. The Voyager mission plan was to fly past Jupiter and Saturn with the aim of returning more images, but this time with new cameras that would allow a higher resolution to be attained not only of the two planets, but also their moons. Voyager 2 was launched on 20 August 1977, followed by Voyager 1 on 5 September 1977.

The images returned revealed that Jupiter had a system of rings, although not by any means as extensive as the Saturnian ring system. Voyager 1 detected considerable volcanic activity on Io, one of Jupiter's moons. A considerable amount of new information about Saturn and its ring system was found by both spacecraft. After its Saturn encounter, Voyager 1 left the ecliptic plane and will not meet another planet. The speed of Voyager 1 is such that, in 1988, it "overtook" Pioneer 10, and became the spacecraft most distant from its home planet.

During the interplanetary cruise phases of their missions, each of the Voyagers use their ultraviolet spectrometers to examine several stellar objects of interest to astronomers.

Voyager 2 is not as healthy as it was. It now has only one working radio receiver (for picking up command signals from Earth) and the scan platform, on which the cameras and ultraviolet spectrometers are mounted, becomes stuck from time to time, especially after it has been moved rapidly. NASA engineers have reprogrammed the onboard computers so that the spacecraft can cope with these two problems, and it is expected that Voyager 2 will perform well during its Neptune encounter in August 1989. In order to make sure the scan platform did not become stuck when it was taking pictures close to Uranus, the computers were told to move the platform slowly. During the close encounter phase, however, the whole spacecraft was made to roll in order to minimize the movements of the scan platform.

Despite the minor technical difficulties with the camera-drive mechanism on Voyager 2, both spacecraft achieved their goals. The trajectory of Voyager 1 meant that after its encounter with Saturn, it would head out of the solar system, climbing out of the ecliptic plane as it did so. However, Voyager 2's post-Saturn trajectory was designed to take it onto fly past Uranus, some four and half years later. Both Voyager craft found new, previously unknown satellites orbiting Jupiter and Saturn. At Uranus, Voyager 2 discovered a staggering ten new satellites in orbit about the planet. The problems with the camera scan platform had been solved and the images of the planet itself, and its retinue of moons, were better than mission planners could have hoped.

In this time exposure, a Titan Centaur rocket blasts off into the night from Kennedy Space Center on December 10, 1974, carrying a Helios 1 space-craft, which approached to within 45 million km. (28 million miles) of the Sun. The Helios programme consists of joint German–US projects.

### Earth-orbiting astronomy satellites

Optical astronomy will make great advances when the Hubble Space Telescope (HST) is launched into orbit. This is due to occur in December 1989, when the Space Shuttle will be used to carry it to orbit some 500 km (310 miles) above the Earth's surface.

HST will be the first long-lived optical observatory in space. Its location above the obscuring and distorting effects of the Earth's atmosphere will provide HST with unique capabilities that will yield a major improvement in observational optical astronomy. It will be able to detect objects seven times further away than we are currently able to see. This means that the volume of the universe available for observation will increase by a staggering 350 times.

The telescope will be equipped with a complement of six scientific instruments, including cameras and spectrographs that will take advantage of HST's unique capabilities. It is expected that some of the instruments will be changed after several years in orbit. HST can be maintained and repaired in orbit, including change of scientific instruments, as necessary, over at least fifteen years. Communications to and from the telescope will take place through the Tracking and Data Relay Satellite System (TDRSS), a geostationary communication satellite system.

The astronomer's end of the HST is the Space Telescope Science Institute (STScI), located on the John Hopkins University's Homewood Campus, in Baltimore, Maryland. Data from the HST will be relayed, via either the TDRS over the Atlantic or the Pacific, to a ground station at White Sands, in New Mexico. From there, the data will be sent to the Space Telescope Operations Control Center (STOCC), in Maryland, via a commercial communications satellite. At the STOCC, the data will be processed and sent by land line to the STScI where the observer may get both the calibrated data as well as the original, raw data – the unprocessed data as it came from the telescope. At times when the telescope is out of contact with both TDRS's, the observations will be stored on-board for later transmission.

At STScI, observers determine the list of targets the telescope will observe, and forward their observing plan to the Science Support Center of the STOCC. Here, the observing plan is converted into an observing schedule which specifies such details as precise observation times, instruments to be used, and instrument observing modes. From the Science Support Center, the observing schedule is passed onto the Payload Operations Control Center (POCC) of the STOCC, which integrates the instrument schedule with other timetables governing the telescope. Such timetables, for example, include the use of the TDRS's, which channel data and instructions between Earth and most NASA low-orbit spacecraft, including the space shuttle, the upcoming space station (Freedom), and the HST.

Next, the staff at the POCC generate the specific instructions enabling the HST to carry out its

SPACE TELESCOPE

observations. At the same time, the operators continue a twenty-four-hour watch on the condition of the telescope's systems. This includes monitoring its electrical power system, attitude control, thermal environment, communications system, and the dozens of other sub-systems necessary for continued, successful operation.

HST consists of a 2·4 m (94 inch) mirror, with an f/24 Cassegrain configuration. The image formed by the main mirror at the focal plane is of extremely high quality. The telescope has been designed to allow, with suitable instrumentation, coverage of the entire wavelength range from the far ultraviolet to the mid-infrared.

Grinding and figuring the 2·4 m (98 inch) primary mirror took some twenty-eight months and the accuracy of it is such that if the mirror was equal to the width of the continental USA then the maximum deviation from the ideal figure would be less than 5 cm (2 inches).

Constructed of low expansion silicate glass, the mirror consists of a honeycomb-like interior sandwiched between two circular plates. The whole structure weighs only 829 kg (1800 pounds) and is 0·3 m (13 inches) thick. The mirror's surface is coated with aluminium to a depth of just 2 millionths of an inch and is protected by a layer of magnesium fluoride 1 millionth of an inch thick.

Electric strip heaters, fixed to the back of both the primary and secondary mirrors, keep their temperature at a constant 21 C to keep their accurate figures.

X-rays simply cannot penetrate the atmosphere to any significant depth and if long-term studies of X-ray sources are required, it is necessary to use Earth orbiting satellites from which to make observations. The first successful X-ray astronomy satellite was UHURU, launched in 1970. This was followed four years later by Ariel V. The data gained by both craft enabled the first detailed maps of the X-ray sky to be produced. The sources discovered included X-ray binaries, clusters of

Observations performed by the Hubble Space Telescope (HST) will be converted to radio waves and transmitted to the ground station at White Sands, New Mexico, via one of the TDRS's. From White Sands, the data will be relayed to the Space Telescope Science Institute at Baltimore for analysis.

galaxies, and Seyfert galaxies. In the second half of the 1970s, the satellites HEAO-1, SAS-3, and Einstein (HEAO-2) were launched. Einstein carried a grazing incidence X-ray telescope which produced a resolution comparable with optical telescopes and provided accurate positions of many X-ray sources. Many of the sources emit 100 to 100,000 times more energy in X-rays than the Sun puts out across the entire spectrum. Others generate bursts of X-rays, or produce them periodically. Many X-ray sources are associated with neutron stars and are candidates for black holes. Gas from a nearby nebula, or companion star, accelerates towards the more massive neutron star and becomes super-hot so that it emits X-rays.

In 1983, the European Exosat was launched. Exosat's task was to measure the intensity of X-rays across a range of energies (or wavelengths) using spectrometers.

Launched in January 1978, the International Ultraviolet Explorer (IUE) was designed for a three-year mission to explore the universe in the ultraviolet region of the spectrum. IUE is still producing very worthwhile results. During its mission IUE has secured high and low resolution spectra of all manner of objects in space, from Halley's Comet to faint, distant quasars.

A potentially serious setback for the mission was the failure of one of the gyroscopes used for pointing and stabilization. Spacecraft engineers at

The Infrared Astronomical Satellite (IRAS) observed the universe from its Sun-synchronous Earth orbit. At the heart of the satellite was a 57 cm super-cooled infrared telescope which revealed previously unsuspected phenomena, both within the Solar System and out to the furthest reaches of space. Electrical power was supplied by the array of solar panels clearly visible in this picture.

NASA's Goddard Spaceflight Center, in Maryland, devised and implemented a new operational mode using the two remaining gyroscopes and the fine Sun sensor. The engineers have already worked out how to control the satellite if one of the two remaining gyros should fail. Measures such as this ensure the longevity of observational programmes.

IUE is a joint US-UK-European project. The IUE Support Team assists astronomers in accessing and using the rich store of data already gained by the satellite. The data store now contains well in excess of 50,000 images.

Because of the success of IUE, a very wide range of topics in observational astrophysics is covered. Research areas include: cometary physics, nearby cool and flaring stars, interstellar dust and gas, emission nebulae, very luminous stars, interacting hot binary stars, cataclysmic variables, the galactic halo, globular clusters, the stellar content of galaxies, very young galaxies, clusters of galaxies, the intergalactic medium, active galactic nuclei (including both nearby Seyfert galaxies and high redshift quasars), and the age and mass of the universe. Many of these projects rely heavily on data acquired by IUE. Nevertheless, the nature of many astronomical objects cannot be determined by looking at one range of wavelengths, and so the IUE data is frequently used in addition to ground-based optical, infrared and radio observations, as well as data gathered from other astronomical satellites, such as IRAS, Exosat and Einstein.

Although ground-based observation can be made through several atmospheric windows up to about 17 micrometres (one micrometre is one millionth of a metre) in the infrared region of the spectrum, measurements at longer wavelengths must be made from balloons, rockets, or satellites. The joint US-UK-Dutch Infrared Astronomical Satellite (IRAS) made some very exciting discoveries from Earth orbit.

Launched in January 1983 from Vandenberg Air Force Base, on a Delta rocket vehicle, IRAS set out on its mission to explore the entire sky in the infrared. It was equipped with a 570 mm (22 inch) Cassegrain telescope, cryogenically cooled to −257 degrees centigrade with liquid helium to detect the faint infrared radiation from clouds of dust in interstellar space. IRAS exhausted its helium coolant in late 1983, but by that time it had gathered, over a period of 300 days, sufficient data to enable a comprehensive map of the universe at infrared wavelengths to be obtained. A catalogue of about a quarter of a million infrared sources has been published.

Since infrared radiation passes more easily than visible light through clouds of dust, IRAS has provided a new view of the structure of the centre of our galaxy, the Milky Way. It has also discovered rings of "dust" particles surrounding various stars – in particular Vega and Beta Pictoris. IRAS did not have the resolving power to determine whether any planets were in orbit about these stars, but future generations of orbiting telescopes may well be able to do so. Subsequent observations of Beta Pictoris, using a 2·5 m (100 inch) telescope at Las Campanas Observatory, in Chile, revealed a flattened disk of dust extending some 60 thousand million kilometres from the star, but again, no planets.

**The future**
What then, of the future? At the time this chapter was being prepared, a number of astronomical satellites and planetary spacecraft were either continuing their missions or being readied for launch. We have already seen that the Hubble Space Telescope will bring about a revolution in astronomy in the visible band as well as the near infrared and ultraviolet bands. Complementary space telescopes will include NASA's Space Infrared Telescope Facility (SIRTF) and the Advanced X-ray Astronomy Facility (AXAF). Both are planned for launch in the 1990s.

Possibly the most ambitious unmanned mission in the next few years will be Ulysses. The prime mission of Ulysses will be to observe the Sun from above one of its poles. The spacecraft will be sent first to Jupiter, to fly over the giant planet's south pole and so gain a gravitational boost which will impart sufficient energy to allow the craft to climb above the ecliptic and years later, to pass several astronomical units over the Sun's north pole.

Undoubtedly, the continued and increased use of unmanned satellites and space probes will help astronomers to unravel the secrets of our universe, and perhaps of our origins. The future then, looks very exciting indeed.

Two major planetary missions are scheduled to begin in 1989. The first will see the Magellan mission to Venus. Magellan is intended to enter a highly elliptical orbit around Venus and use high-resolution radar to map over 90 per cent of the cloud-covered surface. The Magellan radar maps should prove to be the most detailed yet obtained and will allow a far greater understanding of the planet. Around October of the same year, another spaceprobe will set off on its way to the giant planet Jupiter. The Galileo spacecraft consists of two major parts, the orbiter and the descent probe. Both will travel to the vicinity of the Jovian system as a single craft. On the way to Jupiter, Galileo will make measurements of the interplanetary environment and one mission plan calls for a close fly-by of an asteroid. Whether the asteroid fly-by actually occurs has still not been finally decided upon.

## Satellites of Earth and Mars

| number | name | distance from planet (km) | orbital period (days) | size (km) | average magnitude |
|---|---|---|---|---|---|
| Earth | | | | | |
| I | Moon | 384,400 | 27·32 | 3476 | − 12·7 |
| Mars | | | | | |
| I | Phobos | 9350 | 0·30 | 27 × 21 × 19 | 11·6 |
| II | Deimos | 23,490 | 1·26 | 15 × 12 × 11 | 12·7 |

## Satellites of Jupiter

| number | name | distance from planet (km) | orbital period (days) | size (km) | average magnitude |
|---|---|---|---|---|---|
| XVI | Metis | 128,000 | 0·294 | 40 | 17·5 |
| XV | Adrastea | 129,000 | 0·297 | 25 | 18·7 |
| V | Amalthea | 180,900 | 0·489 | 270 × 170 × 150 | 14·1 |
| XIV | Thebe | 221,000 | 0·674 | 80 | 16 |
| I | Io | 421,700 | 1·769 | 3632 | 5·0 |
| II | Europa | 671,000 | 3·551 | 3126 | 5·3 |
| III | Ganymede | 1,070,400 | 7·155 | 5276 | 4·6 |
| IV | Callisto | 1,882,600 | 16·689 | 4820 | 5·6 |
| XIII | Leda | 11,110,000 | 240 | 10 | 20 |
| VI | Himalia | 11,470,000 | 251 | 180 | 14·7 |
| X | Lysithea | 11,710,000 | 260 | 20 | 18·6 |
| VII | Elara | 11,740,000 | 260 | 80 | 16·0 |
| XII | Ananke | 20,700,000 | 617 | 20 | 18·8 |
| XI | Carme | 22,350,000 | 692 | 30 | 18·1 |
| VIII | Pasiphae | 23,300,000 | 735 | 40 | 18·8 |
| IX | Sinope | 23,700,000 | 758 | 30 | 18·3 |

## Satellites of Saturn

| number | name | distance from planet (km) | orbital period (days) | size (km) | average magnitude |
|---|---|---|---|---|---|
| XV | Atlas | 137,000 | 0·602 | 40 × 20 × ? | 18 |
| XVI | Prometheus | 139,400 | 0·613 | 140 × 100 × 80 | 16 |
| XVII | Pandora | 141,700 | 0·629 | 110 × 90 × 70 | 16 |
| X | Janus | 151,400 | 0·694 | 220 × 200 × 160 | 15 |
| XI | Epimetheus | 151,500 | 0·695 | 140 × 120 × 100 | 16 |
| I | Mimas | 186,000 | 0·942 | 392 | 12·1 |
| II | Enceladus | 238,000 | 1·370 | 510 | 11·8 |
| III | Tethys | 295,000 | 1·888 | 1060 | 10·3 |
| XIII | Telesto | 295,000 | 1·888 | 34 × 28 × 26 | 19 |
| XIV | Calypso | 295,000 | 1·888 | 34 × 22 × 22 | 20 |
| IV | Dione | 377,000 | 2·737 | 1120 | 10·4 |
| XII | Helene | 377,000 | 2·737 | 36 × 32 × 30 | 19 |
| V | Rhea | 527,000 | 4·518 | 1530 | 9·7 |
| VI | Titan | 1,222,000 | 15·95 | 5150 | 8·4 |
| VII | Hyperion | 1,481,000 | 21·28 | 410 × 260 × 220 | 14·2 |
| VIII | Iapetus | 3,561,000 | 79·33 | 1460 | 11·0 |
| IX | Phoebe | 12,954,000 | 550 | 220 | 16·5 |

## Satellites of Uranus and Neptune

| number | name | distance from planet (km) | orbital period (days) | size (km) | average magnitude |
|---|---|---|---|---|---|
| Uranus | | | | | |
| 1986U7 | Cordelia | 49,771 | 0·333 | 40? | 22·9 |
| 1986U6 | Ophelia | 53,796 | 0·375 | 50? | 22·6 |
| 1986U9 | Bianca | 59,173 | 0·433 | 50? | 22·6 |
| 1986U3 | Cresida | 61,777 | 0·463 | 60? | 22·2 |
| 1986U6 | Desdemona | 62,676 | 0·475 | 60? | 22·2 |
| 1986U2 | Juliet | 64,352 | 0·492 | 80? | 21·5 |
| 1986U1 | Portia | 66,085 | 0·513 | 80? | 21·5 |
| 1986U4 | Rosalinda | 69,942 | 0·558 | 60? | 22·2 |
| 1986U5 | Belinda | 75,358 | 0·621 | 60? | 22·2 |
| 1985U1 | Puck | 86,000 | 0·763 | 170 | 20·3 |
| V | Miranda | 130,400 | 1·41349 | 485 | 16·5 |
| I | Ariel | 191,700 | 2·520384 | 1160 | 14·4 |
| II | Umbriel | 267,100 | 4·144183 | 1190 | 15·3 |
| III | Titania | 438,300 | 8·705876 | 1610 | 14·0 |
| IV | Oberon | 586,200 | 13·463262 | 1550 | 14·2 |
| Neptune | | | | | |
| I | Triton | 355,200 | 5·876844 | 3500? | 13·5 |
| II | Nereid | 5,562,000 | 359·881 | 300? | 18·7 |

## Satellite of Pluto

| number | name | distance from planet (km) | orbital period (days) | size (km) | average magnitude |
|---|---|---|---|---|---|
| I | Charon | 19,700 | 6·39 | 1200 | 17·0 |

## Asteroid orbits

| number | name | distance from Sun (AU) mean | perihelion | orbit period (years) | size (km) |
|---|---|---|---|---|---|
| 1 | Ceres | 2·7663 | 2·5488 | 4·6012 | 1003 |
| 2 | Pallas | 2·7687 | 2·1136 | 4·6069 | 540 |
| 4 | Vesta | 2·3619 | 2·1528 | 3·6301 | 538 |
| 433 | Eros | 1·4581 | 1·3333 | 1·7607 | 23 |
| 532 | Herculina | 2·7728 | 2·2878 | 4·6173 | 240 |
| 588 | Achilles | 5·2112 | 4·4384 | 11·8964 | 53 |
| 944 | Hidalgo | 5·8201 | 1·9991 | 14·0413 | 16 |
| 1566 | Icarus | 1·0777 | 0·1868 | 1·1188 | 1 |
| 1862 | Apollo | 1·4697 | 0·6468 | 1·78 | ? |
| 2060 | Chiron | 13·6991 | 8·5126 | 50·70 | 150–650? |
| 1973 NA | | 2·4470 | 0·8796 | 3·83 | ? |
| 1975 YA | | 1·2901 | 0·9054 | 1·47 | ? |
| 1976 AA | | 0·9664 | 0·7899 | 0·95 | 1 |
| 1976 UA | | 0·8440 | 0·4643 | 0·76 | 0·3? |

## Messier objects

| Messier number | constellation | position RA hr min. | Dec. deg. min. | visual magnitude | name or type of object |
|---|---|---|---|---|---|
| 1 | Tau | 05 31·5 | +21 59 | 8·4 | Crab Nebula (supernova remnant) |
| 2 | Aqr | 21 30·9 | −01 03 | 6·3 | Globular Cluster |
| 3 | CVn | 13 39·9 | +28 38 | 6·2 | Globular Cluster |
| 4 | Sco | 16 20·6 | −26 24 | 6·1 | Globular Cluster |
| 5 | Ser | 15 16·0 | +02 16 | 6·0 | Globular Cluster |
| 6 | Sco | 17 36·8 | −32 11 | 5·5 | Open Cluster |
| 7 | Sco | 17 50·7 | −34 48 | 5 | Open Cluster |
| 8 | Sgr | 18 01·6 | −24 20 | 5·8 | Lagoon Nebula |
| 9 | Oph | 17 16·2 | −18 28 | 7·6 | Globular Cluster |
| 10 | Oph | 16 54·5 | −04 02 | 6·4 | Globular Cluster |
| 11 | Sct | 18 48·4 | −06 20 | 6·5 | Open Cluster |
| 12 | Oph | 16 44·6 | −01 52 | 6·7 | Globular Cluster |
| 13 | Her | 16 39·9 | +36 33 | 5·8 | Globular Cluster |
| 14 | Oph | 17 35·0 | −03 13 | 7·8 | Globular Cluster |
| 15 | Peg | 21 27·6 | +11 57 | 6·3 | Globular Cluster |
| 16 | Ser | 18 16·0 | −13 48 | 6·5 | Open Cluster |
| 17 | Sgr | 18 18·0 | −16 12 | 7 | Omega Cluster |
| 18 | Sgr | 18 17·0 | −17 09 | 7·2 | Open Cluster |
| 19 | Oph | 16 59·5 | −26 11 | 6·9 | Globular Cluster |
| 20 | Sgr | 17 58·9 | −23 02 | 8·5 | Trifid Nebula |
| 21 | Sgr | 18 01·8 | −22 30 | 6·5 | Open Cluster |
| 22 | Sgr | 18 33·3 | −23 58 | 5·3 | Globular Cluster |
| 23 | Sgr | 17 54·0 | −19 01 | 6·5 | Open Cluster |
| 24 | Sgr | 18 15·5 | −18 27 | 5 | Open Cluster |
| 25 | Sgr | 18 28·8 | −19 17 | 6 | Open Cluster |
| 26 | Sct | 18 42·5 | −09 27 | 9·1 | Open Cluster |
| 27 | Vul | 19 57·4 | +22 35 | 8·1 | Dumbell Nebula (planetary nebula) |
| 28 | Sgr | 18 21·5 | −24 54 | 7·1 | Globular Cluster |
| 29 | Cyg | 20 22·2 | +38 21 | 7·2 | Open Cluster |
| 30 | Cap | 21 37·5 | −23 25 | 7·7 | Globular Cluster |
| 31 | And | 00 40·0 | +41 00 | 4·0 | Andromeda Galaxy (Spiral) |
| 32 | And | 00 40·0 | +40 36 | 8·5 | Elliptical Galaxy |
| 33 | Tri | 01 31·1 | +30 24 | 6·0 | Spiral Galaxy |
| 34 | Per | 02 38·8 | +42 34 | 5·7 | Open Cluster |
| 35 | Gem | 06 05·7 | +24 20 | 5·6 | Open Cluster |
| 36 | Aur | 05 32·0 | +34 07 | 6·0 | Open Cluster |
| 37 | Aur | 05 49·0 | +32 23 | 6·0 | Open Cluster |
| 38 | Aur | 05 25·3 | +35 48 | 7 | Open Cluster |
| 39 | Cyg | 21 30·4 | +48 13 | 5 | Open Cluster |
| 40 | UMa | 12 33·0 | +58 30 | | Double Star |
| 41 | CMa | 06 44·9 | −20 42 | 5 | Open Cluster |
| 42 | Ori | 05 32·9 | −05 25 | 4 | Orion Nebula |
| 43 | Ori | 05 33·1 | −05 18 | 9 | Orion Nebula |
| 44 | Cnc | 08 37·5 | +19 52 | 3·7 | Praesepe (Open Cluster) |
| 45 | Tau | 03 43·9 | +23 58 | 1·6 | Pleiades (Open Cluster) |
| 46 | Pup | 07 39·6 | −14 42 | 6 | Open Cluster |
| 47 | Pup | 07 34·3 | −14 22 | 5 | Open Cluster |
| 48 | Hya | 08 11·3 | −05 39 | 6 | Open Cluster |
| 49 | Vir | 12 27·3 | +08 16 | 8·9 | Elliptical Galaxy |
| 50 | Mon | 07 00·5 | −08 16 | 6·5 | Open Cluster |
| 51 | CVn | 13 27·8 | +47 27 | 8·4 | Whirlpool Galaxy (spiral) |
| 52 | Cas | 23 22·0 | +61 20 | 7·1 | Open Cluster |
| 53 | Com | 13 10·5 | +18 26 | 7·7 | Globular Cluster |
| 54 | Sgr | 18 52·0 | −30 32 | 7·7 | Globular Cluster |
| 55 | Sgr | 19 36·9 | −31 03 | 6·1 | Globular Cluster |
| 56 | Lyr | 19 14·6 | +30 05 | 8·3 | Globular Cluster |
| 57 | Lyr | 18 51·7 | +32 58 | 9·0 | Ring Nebula (planetary nebula) |
| 58 | Vir | 12 35·1 | +12 05 | 9·9 | Barred Spiral Galaxy |
| 59 | Vir | 12 39·5 | +11 55 | 10·2 | Elliptical Galaxy |
| 60 | Vir | 12 41·1 | +11 48 | 9·2 | Elliptical Galaxy |
| 61 | Vir | 12 19·4 | +04 45 | 9·8 | Barred Spiral Galaxy |
| 62 | Sco | 16 58·1 | −30 03 | 7·1 | Globular Cluster |
| 63 | CVn | 13 13·5 | +42 17 | 8·9 | Spiral Galaxy |
| 64 | Com | 12 54·3 | +21 47 | 8·7 | Spiral Galaxy |
| 65 | Leo | 11 16·3 | +13 23 | 9·6 | Spiral Galaxy |

## Messier objects

| Messier number | constel-lation | RA hr min. | Dec. deg. min. | visual mag-nitude | name or type of object |
|---|---|---|---|---|---|
| 66 | Leo | 11 17·6 | +13 17 | 9·1 | Spiral Galaxy |
| 67 | Cnc | 08 48·3 | +12 00 | 6·3 | Open Cluster |
| 68 | Hya | 12 36·8 | −26 29 | 8·0 | Globular Cluster |
| 69 | Sgr | 18 28·1 | −32 23 | 7·8 | Globular Cluster |
| 70 | Sgr | 18 40·0 | −32 21 | 8·3 | Globular Cluster |
| 71 | Sge | 19 51·5 | +18 39 | 7·5 | Globular Cluster |
| 72 | Aqr | 20 50·7 | −12 44 | 9·2 | Globular Cluster |
| 73 | Aqr | 20 56·4 | −12 50 | | Open Cluster |
| 74 | Psc | 01 34·0 | +15 32 | 9·6 | Spiral Galaxy |
| 75 | Sgr | 20 03·2 | −22 04 | 8·3 | Globular Cluster |
| 76 | Per | 01 38·8 | +51 19 | 11·5 | Planetary Nebula |
| 77 | Cet | 02 40·1 | −00 14 | 9·1 | Spiral Galaxy (Seyfert type) |
| 78 | Ori | 05 44·2 | +00 02 | | Small Emission Galaxy |
| 79 | Lep | 05 22·2 | −24 34 | 7·4 | Globular Cluster |
| 80 | Sco | 16 14·1 | −22 52 | 7·2 | Globular Cluster |
| 81 | UMa | 09 51·5 | +69 18 | 7·0 | Spiral Galaxy |
| 82 | UMa | 09 51·9 | +69 56 | 8·7 | Irregular Galaxy |
| 83 | Hya | 13 34·3 | −29 37 | 7·6 | Spiral Galaxy |
| 84 | Vir | 12 22·6 | +13 10 | 9·7 | Elliptical Galaxy |
| 85 | Com | 12 22·8 | +18 28 | 9·5 | Elliptical Galaxy |
| 86 | Vir | 12 23·7 | +13 13 | 9·8 | Elliptical Galaxy |
| 87 | Vir | 12 28·3 | +12 40 | 9·3 | Elliptical Galaxy |
| 88 | Com | 12 29·5 | +14 42 | 9·8 | Spiral Galaxy |
| 89 | Vir | 12 33·1 | +12 50 | 10·2 | Elliptical Galaxy |
| 90 | Vir | 12 34·3 | +13 26 | 9·7 | Spiral Galaxy |
| 91 | Com | 12 34·0 | +11 32 | 10·3 | Barred Spiral Galaxy |
| 92 | Her | 17 15·6 | +43 12 | 6·3 | Globular Cluster |
| 93 | Pup | 07 42·4 | −23 45 | 6 | Open Cluster |
| 94 | CVn | 12 48·6 | +41 23 | 8·1 | Spiral Galaxy |
| 95 | Leo | 10 41·3 | +11 58 | 9·9 | Barred Spiral Galaxy |
| 96 | Leo | 10 44·2 | +12 05 | 9·4 | Spiral Galaxy |
| 97 | UMa | 11 12·0 | +55 18 | 11·2 | Owl Nebula (Planetary Nebula) |
| 98 | Com | 12 11·3 | +15 11 | 10·4 | Spiral Galaxy |
| 99 | Com | 12 16·3 | +14 42 | 9·9 | Spiral Galaxy |
| 100 | Com | 12 20·4 | +16 06 | 9·8 | Spiral Galaxy |
| 101 | UMa | 14 01·4 | +54 35 | 8·2 | Spiral Galaxy |
| 102 | Dra | 15 05·1 | +55 57 | 10·5 | Highly Included Spiral Galaxy |
| 103 | Cas | 01 29·9 | +60 27 | 7 | Open Cluster |
| 104 | Vir | 12 37·3 | −11 21 | 8 | Sombrero Nebula |
| 105 | Leo | 10 45·2 | +12 51 | 9·5 | Elliptical Galaxy |
| 106 | CVn | 12 16·5 | +47 35 | 9 | Spiral Galaxy |
| 107 | Oph | 16 29·7 | −12 57 | 9 | Globular Cluster |
| 108 | UMa | 11 08·7 | +55 57 | 10·5 | Spiral Galaxy |
| 109 | UMa | 11 55·0 | +53 39 | 10·6 | Barred Spiral Galaxy |
| 110 | And | 00 37·6 | +41 25 | − | Small Elliptical Galaxy |

## Bright globular clusters

| cluster name | RA hr min. | Dec deg. min. | visual mag. | distance (light-years) |
|---|---|---|---|---|
| 47 Tucanae | 00 21·9 | −72 21 | 4·0 | 16,300 |
| ω Centauri | 13 23·8 | −47 03 | 3·6 | 17,000 |
| M3 | 13 39·9 | 28 38 | 6·4 | 34,500 |
| M5 | 15 16·0 | 02 16 | 5·9 | 26,400 |
| M4 | 16 20·6 | −26 24 | 5·9 | 14,000 |
| M13 | 16 39·9 | 36 33 | 5·9 | 20,500 |
| M92 | 17 15·6 | 43 12 | 6·1 | 25,700 |
| M22 | 18 33·3 | −23 58 | 5·1 | 9800 |
| Δ 295 | 19 06·4 | −60 04 | 6·2 | 17,300 |
| M15 | 21 27·6 | 11 57 | 6·4 | 34,200 |
| M2 | 21 30·9 | −01 03 | 6·3 | 40,100 |

## Bright star clusters

| | RA hr min. | Dec deg. min. | visual mag. | distance (light-years) |
|---|---|---|---|---|
| h & χ Persei | 02 15·5 | 56 55 | 4·2 | 7700 |
| M34 | 02 38·8 | 42 34 | 5·6 | 1400 |
| M45 (Pleiades) | 03 44·1 | 23 57 | 1·3 | 410 |
| Hyades | 04 17 | 15 30 | 0·6 | 147 |
| M38 | 05 25·3 | 35 48 | 7·0 | 4300 |
| M36 | 05 32·8 | 34 06 | 6·3 | 4100 |
| M37 | 05 49·1 | 32 32 | 6·1 | 4170 |
| M35 | 06 05·8 | 24 21 | 5·3 | 2840 |
| M44 (Praesepe) | 08 37·2 | 20 10 | 3·7 | 515 |
| M67 | 08 47·8 | 12 00 | 6·5 | 2700 |
| κ Crucis (Jewelbox) | 12 50·7 | −60 04 | 5·0 | 2700 |
| M21 | 18 01·6 | −22 30 | 6·8 | 4080 |
| M16 | 18 16·0 | −13 48 | 6·6 | 8150 |
| M11 | 18 48·4 | −06 20 | 6·3 | 5670 |
| M39 | 21 30·4 | 48 13 | 5·1 | 815 |

# Index